EMBO

MW01598768

Critical Food Studies

Series Editor
Michael K. Goodman, Kings College London, UK

The study of food has seldom been more pressing or prescient. From the intensifying globalization of food, a world-wide food crisis and the continuing inequalities of its production and consumption, to food's exploding media presence, and its growing re-connections to places and people through 'alternative food movements', this series promotes critical explorations of contemporary food cultures and politics. Building on previous but disparate scholarship, its overall aims are to develop innovative and theoretical lenses and empirical material in order to contribute to – but also begin to more fully delineate – the confines and confluences of an agenda of critical food research and writing.

Of particular concern are original theoretical and empirical treatments of the materializations of food politics, meanings and representations, the shifting political economies and ecologies of food production and consumption and the growing transgressions between alternative and corporatist food networks.

Other titles in the series include:

Liquid Materialities
A History of Milk, Science and the Law
Peter Atkins
9780754679219

Embodied Food Politics

MICHAEL S. CAROLAN
Colorado State University, USA

Routledge
Taylor & Francis Group

LONDON AND NEW YORK

First published 2011 by Ashgate Publishing

2 Park Square, Milton Park, Abingdon, Oxon OX14 4RN
711 Third Avenue, New York, NY 10017, USA

Routledge is an imprint of the Taylor & Francis Group, an informa business

First issued in paperback 2016

British Library Cataloguing in Publication Data
Carolan, Michael S.
  Embodied food politics. -- (Critical food studies)
  1. Agriculture--Social aspects. 2. Food industry and
  trade--Social aspects.
  I. Title II. Series
  306.3'49-dc22

Library of Congress Cataloging-in-Publication Data
Carolan, Michael S.
  Embodied food politics / by Michael S. Carolan.
    p. cm. -- (Critical food studies)
  Includes bibliographical references and index.
  ISBN 978-1-4094-2209-9 (hardback) -- ISBN 978-1-4094-2210-5 (ebook)
  1. Food habits--Political aspects. 2. Food preferences--Political aspects. 3. Food industry and trade--Political aspects. I. Title.
  GT2850.C36 2011
  394.1'2--dc22

2010046226

ISBN 978-1-4094-2209-9 (hbk)
ISBN 978-1-138-24575-4 (pbk)

Transferred to Digital Printing 2014

Contents

List of Figures and Tables

Figures

Tables

Acknowledgments

If I am to be true to my own words, I have a lot of people to be thankful for.[1] "Being with" these people has provided me with embodiments that have carried me to, and continue to carry through, this moment. My parents, for example, who haunt this book in various places, exposed me to foodscapes that shape my *feel* of food—and, I would argue, even agro-food scholarship—to this day. Without you I can't imagine thinking about food, as either a scholar or eater, as I do now.

I have also had the benefit of thinking through many of the ideas discussed in this book in the settings of professional conferences, public lectures, and the classroom. To all those who have discussed these issues with me, whether in person or from some distant computer: thanks. Thanks also to all those who were kind enough to participate in the research that makes this book possible. You've all taught me more than you know. Sharing your understandings of the world with me has brought this book to life. Without you, the scholarship I attempt would sound hollow, feel cold—it would be, in a word, dead.

While more than 70 individuals give life to this manuscript one is responsible for getting it published: Michael Goodman. That we're both making a "visceral turn" at the same time; I don't know what else to call it other than serendipity. This shared interest not only made me a friend but also opened a door and allowed me to obtain a book contract with Ashgate Publishing in an exciting new series titled Critical Food Studies. Thanks Michael (and Ashgate) for taking a chance on a manuscript with an admittedly unusual conceptual approach to issues of food consumption and production. Thanks also to Debbie Carolan, graphic designer (and sister) extraordinaire. Recognizing the importance of first impressions, I owe you a debt of thanks for making this book look, at first glance, really interesting. Debbie is responsible for the book's captivating cover image.

Finally: Nora. This "being with" I could least be without. You enrich my grip upon the world in so many ways. For everything: thank you.

---

1   While I am happy to share the stage when it comes to what's right, interesting, and/or valuable about this book I am unwilling to allocate blame due to errors and/or omissions to anyone but myself.

Chapter 1

Thinking About Food Relationally

Canned mushrooms … I loved them as a kid. Their saltiness and texture made for a delicious combination. I have a memory—I must have been about six—of going to a pizzeria while on a trip with my parents and sister. Like always, we ordered a pizza with a variety of toppings, including mushrooms. When it arrived I'm sure I demanded the first piece and hastily took a bite. My taste buds betrayed all expectations. With disgust I asked my parents about the oddly shaped, offensively tasting objects in front of me masquerading as mushrooms. They're fresh mushrooms, I remember them saying. It was not until University that I ate fresh mushrooms again. I grew up, you see, in a small town (about 350 people) in northeastern Iowa. Fresh mushrooms were not available in our local grocery store, though they are today. That was then. Today I enjoy fresh mushrooms even more than canned.

I share this story with others, usually in the setting of the kitchen with someone leaning over a chopping board slicing these delicious fungi. I am then typically asked to account for this transformation. I guess I just didn't know better, was my standard response. I did not give that response much thought until recently.

I have long been fascinated by the subjects of knowledge, cognition, and perception, particularly from the angle of embodiment. What I have learned, and my own research has only strengthened my convictions, is that understandings of the world are inextricably shaped by *lived* experience. There is no mind, no mind's eye or disembodied reason, at least as those "things" are traditionally conceived in the Anglo-American analytic tradition that places a premium on a free and autonomous rational faculty. Rather, what we call mind, thought, cognition, and knowledge are all *effects* of active bodies, of bodies-in-the-world. Andy Clark (2008) uses the term "supersizing the mind" to emphasize this smearing across mind, body and environment as we attempt to make sense—literally—of the world. Others speak of the embodied mind (Varela et al. 1992). Social scientists—because they too are interested in this phenomenon—evoke terms like dwelling, skill, practice, and non-representational when speaking of the supra-mental, supra-discursive processes that underlie knowledge (see e.g., Bourdieu 1995; Ingold 2000; Thrift 2004). Call it whatever you want, the underlying point throughout these literatures remains the same: we think with and through our bodies.

Some of my own recent research has highlighted how lived experience shapes peoples' understanding of phenomena. Specifically, I have documented how different embodiments produce divergent understandings of things like "nature" (Carolan 2007, 2009a) and "countryside" (Carolan 2008). What, then, about our understandings of food? How do our lived experiences, our material practices, and our sensorial engagements shape how we think about issues related to food consumption and production?

Back to mushrooms. Traditional understandings of knowledge would view my explanation for why I once preferred canned over fresh—I guess I just didn't know better—as an admittance of ignorance. If someone would have only told me that fresh mushrooms taste better than canned, or if I read this somewhere, then that memory from so long ago would have unfolded differently. Of course this is nonsense. I had to learn to like them. But what does it mean to say I have *learned* to like fresh mushrooms, since clearly this knowledge has been acquired through means other than reading books or listening to others? I am referring to a type of learning that is active and engaged, which involved (and still involves) not only eating fresh mushrooms but cooking with them, sautéing them (just writing this evokes the sounds and smells of mushrooms crackling in a sauté of garlic and butter), and eating them in the company of friends and family (food scholar Sidney Mintz [1996: 98], for example, argues that the foods we most appreciate are made through collective relationships).

I am reminded of something the philosopher Ludwig Wittgenstein (1969: number 467) once wrote:

> I am sitting with a philosopher in the garden; he says again and again "I know that's a tree", pointing to a tree that is near us. Someone else arrives and hears this, and I tell him: "This fellow isn't insane. We are only doing philosophy."

What Wittgenstein's friend failed to grasp is how their lived experience of the world—a world populated by trees—played into their knowing of these phenomena (see also, Harrison 2000: 507). What this individual missed is that our understanding of trees is wrapped up in non-linguistic forms of knowledge that come about through our *doings* with all things wooden—such as taking walks in the forest, smelling freshly cut wood, dancing on hardwood floors, getting splinters, and listening to the rustling of the leaves on a cool fall afternoon. The same, I contend, pertains to how we know food.

Let us now extrapolate out from this. What happens to our understanding of "food quality" as we become increasingly conditioned to eating and cooking with industrialized food, which places a premium on making sure its products are standardized and fit for the rigors of international travel? How are our understandings of, say, apples or beef shaped by our lived experiences of these phenomena? Equally important are questions related to the knowledges disappearing as a result of vanishing material practices. As food has become standardized and agriculture further grounded in techniques of monocropping think of all the embodied experiences lost. Today, for example, our understanding of most fruits and vegetables is limited to just a handful of varieties. Corn, tomatoes, and apples look, taste, smell, and feel little today like when Black Aztec (sweetcorn), Big Rainbows (tomatoes), and the prized Rambo (apple) populated gardens, backyards, dinner tables, and stomachs.

The relational basis of our knowledge of food has long been acknowledged. Studies dating back to the late nineteenth century showed the effects that urbanization was having on children in terms of their food knowledge. A study conducted in 1880 on 200 middle-class Boston first-graders showed the clear effects that these new urban embodiments were having on how people understood what they ate. Almost 90 percent of those surveyed did not understand the fundamentals of a wheat field, 75 percent did not know about the seasons, and "more than 60 percent had no concept of a beehive, a crow, a robin, or a bluebird, or of planting seeds or growing beans, potatoes, strawberries, blueberries, blackberries, or corn" (Vileisis 2007: 104). About once a year a study like this is mentioned in the press, making the point of how few children know that milk comes from cows or that pickles are cucumbers. Had their lived experiences been different, to include encounters with dairy parlors and "pickle days" (what my mom calls days when pickles are made), I have no doubt the children surveyed would have responded differently. More often than not these survey findings are merely a source of indignation, evidence that children need to be better educated about food and the food system more generally. But the "education" often advocated is incomplete; too centered on what is called representational knowledge—on pictures, words, diagrams, and lectures. This solution, however, is short-sighted. As the encounter between Wittgenstein and his friend illustrates, the meanings of phenomena—whether a tree or a food system—cannot be conveyed with a finger point, some words, and a picture.

Agro-food scholars have in the last decade taken a relational approach to studying the food system (Goodman 2001). Here, the concrete practices of everyday life are being draw into the methodological horizon of the analyst (see for example DeLind 2006; Goodman et al. 2010; Roe 2006; Lockie 2002; Whatmore 2002). Traditional commodity chain analyses seek to "follow the commodity" (de Sousa and Busch 1998: 252) from farm-to-fork (see also Goodman and Dupuis 2002: 6–7). The relational approach, conversely, redirects the "myopic productionist 'gaze' that has dominated much of agro-food studies work" (Roe 2006: 106; see also Lockie and Kitto 2000: 15; Lockie 2002: 290; Goodman 2004: 13), focusing instead upon the lived world of meaning generation (see Lockie and Kitto 2000: 13–15; Roe 2006: 113–17). This shift in analytic focus reveals the transformational potential that lies at the site of the fork. It is this site that interests me, where "things become food" (Roe 2006: 107) and food, sometimes, becomes a thing.

It would be unfair, however, to classify this book as being entirely about food consumption. Goodman and DuPuis (2002: 11) made an important observation a few years back that still largely holds true today, noting how consumption-focused food studies too often neglect the "production side of food". Granted, social scientists make analytic "cuts" all the time, otherwise it would be nearly impossible to talk about anything (because we would have to talk about everything). But we need to be aware of what we miss when one "side" is examined at the expense of the other. Goodman and DuPuis (2002: 11) write about wanting to "build a better theoretical bridge between [historically consumption-oriented] food studies and [historically production-oriented] agro-food studies". My goals are more modest, though I believe the approach adopted here offers a sincere attempt at erecting some theoretical and empirical connections to these otherwise opposing "ends". Given that living bodies populate our food system it makes sense that an approach examining embodied experience would have something to say about what we eat, regardless of whether those bodies happen to be analytically lumped into the category of "producer" or "consumer".

Learning to be Affected

"Pure" experience is a chimera. After penning one of the most familiar sentences in Western thought, "I think therefore I am", Descartes should have added "... and what I am—my lived experience—is always-already folded into what and how I think". You see, there is no singular consciousness; no one way to reach out and know the world sensually. Our knowledge of the world, rather, is constituted through our relationalities. And as those relationalities change so too change understandings of what is and what ought to be.

A few years ago Bruno Latour (2004) wrote an article on the training of "noses" for the perfume industry. Latour describes to the reader what is known in the industry as an odor kit. The kit consists of a series of fragrances, arranged from stark to subtle contrasts. With the help of this kit and other "tuned" bodies who already embody the necessary odor sensibilities individuals acquire a "nose" for previously unrecognizable differences in scent. What's interesting about this case, according to Latour, is not its exceptionality. Just the opposite: to be a body is to be a body in articulation with other objects, technologies, spaces, and subjects. Being, to put it in philosophical terms, is all about "learning to be affected" (Latour 2004a: 206). To be an unaffected body is the same as being dead. Latour uses the example of the trained "nose" to force the reader to revisit conventional wisdom about it means to *do* science. In doing this he challenges conventional wisdom that the scientist represents an object-ive (non relational) body *par excellence*. In his own words:

> [I]f I, a tuned nose, need the odour kit to become sensitive to contrasts , chemists need their analytical instruments to render themselves sensitive to differences of one single displaced atom. They too acquire a body, a nose, an organ, through their laboratories this time, and also thanks to their conferences, their literature and all the paraphernalia that make up what could be called the *collective body* of science (Latour 2004a: 209).

Latour's point is that scientists are not unique when it comes to knowing the world; knowing—even scientific ways of knowing—presupposes bodies that have learned to be affected. It therefore follows that if knowing the world is in part a product of learning to be affected by that world the same should hold for our knowing of what we eat. Like the trained "nose" found in the perfume industry we too are tuned: to food.

A few years back I had a friend visit my home in Fort Collins, Colorado (for those unfamiliar with the town, Fort Collins is nestled against the foothills of the Rocky Maintains). Walking around town I remember him looking west to the mountains and proclaiming: "I don't care where you're from, those are simply breathtaking". Allow me to quickly unpack this assertion in the light of what has just been said; after which I'll get to the subject of food.

Rather than breathtaking, mountains during the Middle Ages were widely viewed as ominous and dangerous. Among French farmers, for instance, certain "dangerous mountains" were known as unfit for pasture—at least until the advent of certain vaccinations—as sheep that grazed on them contracted anthrax (Dubos 2007: 273). Aesthetically speaking mountains fared little better, being described in the seventeenth-century literature world as "Earth's Dugs, Risings, Tumors, Blisters, Warts" (Tuan 1990: 72). For our ancestors, wilderness was, quite simply, too wild to be looked at through the romantic lens that many today view this space through. Microbiologist, environmental historian, and Pulitzer Prize winner René Dubos (1980:14) provides the following account of these spaces of wilder-ness:

> Until the eighteenth century [...] the Derbyshire peak region in England was considered wild and unfit for human eyes. In 1681, the poet Charles Cotton described it as a 'country so deformed' that it might be regarded as "Nature's pudenda." Travelers in those days were advised to keep their coach blinds drawn while traversing the region so as not to be shocked by its ugliness and wildness.

My friend's positive understanding of the mountains is a privileged effect. The weather at 10,000 feet can change in a matter of minutes. I can only image the trepidation early travelers must have felt as they slowly made their way through a mountain pass in their un-heated/air-conditioned carriage, lacking the life-line of a cell phone and satellite navigation system and with little insight into what the next hour would bring in terms of the weather. While I frequently tell my friends about the wonderful summers along the Front Range and in Fort Collins in particular I know this experience is a socio-material effect. The ubiquity of well air-conditioned buildings (and cars) has kept me from knowing a Colorado summer like that experienced by earlier generations. Massive public works projects that now supply water to this area, for agricultural irrigation, residential sprinkler systems, and recreation (such as Horsetooth Reservoir that borders Fort Collins to the west), keep the crops growing, the grass green, and the residents entertained. This is not to suggest that previous generations did not—or could not—view Colorado summers in a positive light. According to historical accounts, many in fact did. My point, rather, is that understandings of this space—like understandings of *anything*—are inextricably a socio-technically mediated effect.

We cannot know this space *except* through our relations with it. And the same, I would contend, holds for our understandings of food.

Thinking about our knowledge and understanding of food and the food system more generally in this manner—that is, as something that we *do* rather than as something that we objectively acquire—radically alters the parameters of the debate surrounding food production and consumption. For instance, I routinely come across arguments that speak of how "[w]e are hardwired to love the taste of fat, salt, and sugar" (Wansink 2007: 180). This biological fact, which I have no reason to dispute, is then used to argue that our attraction to fast food is both inevitable and natural—after all, it's rooted in our genes (see also Allman 1995: 50; Cartwright 2000: 47). Not only does this essentialize taste but it also washes away the initial visceral resistant put up by bodies that had not yet learned to be affected by industrial food. Anthropologist Melissa Caldwell (2004: 15), for example, when studying the introduction of McDonald's food in Moscow, reports that many respondents initially did not like its taste. One individual went as far to explain "that he had tried it and could not understand why a person would eat such food more than to try it once" (p. 15).

To evoke the words of Latour, Muscovites (like me with fresh mushrooms) had to learn to be affected by McDonald's and its food. The Ronald McDonald's army is made, not born. This in itself is a very important point. If we recognize that it took work to make bodies tuned to fast food (and industrial food more generally) then we must recognize that the reverse is also true—that there must be, if you will, a "re-tuning" toward alternative foods and food systems if we want those alternatives to be sustained over the long run. It therefore comes as no surprise to me when I hear school administrators talk about how students frequently "choose" the less healthy meal when given the option of, say, French fries or a salad. That "choice", it turns out, is just one point of a whole series of material connectivities; connectivities that otherwise go ignored when all one looks at is the moment of purchase. As documented in the following chapters, so called consumer "preference" is produced and maintained through practice—through literally *doing* those tastes over and over again. It's insincere to say we ought to let those preferences alone dictate what food is provided and how that food is produced when today's dominant preferences were produced through non market means.

The relational approach adopted here also says something about food politics: that removing the "choice" of McDonald's French fries from school cafeterias, eating whole foods, and even growing one's own food are all profoundly political acts. Political theory often speaks of politics as involving only subjects. The vulgarities of materiality—of emotion, impulse and other corporeal activities—have no place, it has long been argued, in reasoned discourse. Along these lines, Hannah Arendt (1958: 80–89) makes a distinction between "action" and "work". The former refers to the turn-taking discourse we often ascribe to (ideal) political debate. Work, in contrast, involves those activities one engages in for survival. Arendt believed political debate ought to be insulated as much as possible from the material realities of everyday life so as to minimize self-interested behavior. Jürgen Habermas (1984: 86) makes a similar distinction in his writings on communicative action, in which actors in society seek to reach common understanding by reasoned argument, consensus, and cooperation rather than through action strictly in pursuit of their own goals. Yet this view clings too stubbornly to Descartes' legacy, in that mind—the prime mover in Western cosmology—is privileged over the "dead" realm of the material (Law and Mol 2008: 134–6).

The problem with this conventional understanding of politics is that it is premised upon a world that does not exist. There is no purely rational faculty divorced from the materialities of everyday life. Given that the world is smeared—or populated by "hybrids" (Latour), "monsters" (Law), and "Cyborgs" (Haraway)—we had better get used to talking about how *relations* have politics too. For if our understandings of what is and what ought to be are shaped by our material connectivities, then any change in those connections ought to have an effect on how we view the world. A relational (food) politics thus avoids the thorny problem faced by analyses grounded in a more conventional view of politics and the liberal humanism therein presupposed in that it actually *can* say something about why we should care as well as being instructive in cases where we don't (Braidotti 2006: 119; McEwan and Goodman 2010: 110).

Situating the Book Conceptually and Empirically

This book seeks to make conceptual and empirical contributions to the field of agro-food studies by entering into the subjects of food and food politics by way of the lived experience. In doing this, I hope to provide the reader with an understanding of the grip on food nurtured through community supported agriculture (known commonly as CSA), heritage seed banks, and backyard chicken coops. My choice of the term "grip" is intentional. As used by the Maurice Merleau-Ponty (see for instance, 1992: 289–90), grip is a metaphor to convey how we know the world. It speaks to knowing as a richly socio-historical, thoroughly sensuous experience that does not just happen but which involves constant bodily adjustments as we seek its maximization. Clasp your hands together. Notice how you cannot tell where one grip ends and the other begins. The same holds for perception in general. The knower and the known constitute the "understanding" produced. We too are clasped with the world. And within this embrace lies our knowledge of food.

So what competencies, knowledges, and sentiments do the embodied practices cultivated in these three empirical spaces bring forth? Answering this question reveals, among other things, the transformational potential of phenomena like CSA, heritage seed banks, and backyard chicken coops. Global Food (shorthand for a system of large scale, global food provisioning), through the embodiments it creates, helps foster particular knowledges, tastes, and feelings about food. These understandings (and the consumer "preferences" they enact) give support to conventional food production and consumption. One thread that the following chapters share is that CSA, heritage seed banks, and backyard chicken coops threaten the conventional system by introducing competencies, knowledges, and sentiments that make problematic the artifacts, practices, and visceral experiences of Global Food. These spaces, quite literally, help to make Global Food out of tune to bodies that come into repeated contact with them.

Unlike most empirically grounded books, which are often centered on a single case study, the aforementioned goals are pursued through a handful of cases, three to be exact. The rationale for this is simple: if our knowledge of food is relational then different relations have different epistemic effects. Examining three distinct spaces, and the socio-material relations therein embedded, allows me to better detail the lived experience they provide while exploring how those experiences create friction for Global Food. More is produced than just food or seed in these cases. It is this "more than" that I am interested in, as well as how this "more than" differs from that nurtured by Global Food. I will now elaborate on the concept of the lived experience.

While I do my best in this book to avoid coining new neologisms, recasting this term in this case is necessary. That is because I am not entirely faithful to any one theoretical tradition that claims to deal explicitly with the phenomena of embodied knowledge. Nonrepresentational Theory (see e.g., Latham 2003; Thrift 2004) and feminist geography (see e.g., Hayes-Conroy and Hayes-Conroy 2008; Probyn 1999) represent two popular research strands in the social sciences that articulate, in different ways, why and how bodies matter. And I draw from both in the following chapters.

Nonrepresentational Theory finds significance in the insignificant, in things like fleeting encounters, everyday routines, embodied movements, and affective intensities. According to this framework, the "deadening effect" of conventional social scientific analysis, where meaning is viewed as something *imposed* upon an otherwise passive world, "can […] be overcome by allowing in much more of the excessive and transient aspects of living" (Lorimer 2005: 83). The goal of this approach, as with feminist geography, is to move beyond the Western dualisms that house mind and body in separate worlds. Yet we must be careful not to throw out the baby of personal experience with the bathwater of dualistic cosmology. This gets at one problem with Nonrepresentational Theory, at least some flavors of it; a criticism also echoed by feminist scholars (see e.g., Bondi 2005; Hayes-Conroy and Hayes-Conroy 2008; Thein 2005). The critique is that Nonrepresentational Theory goes too far in its exploration of the transient. This removes selfhood as an event-*full* analytic construct and empties personal experience and biography as meaningful theoretical categories. This might explain why Nonrepresentational Theory often does not get us much further than theory. To be sure, there are some honorable attempts to put empirical flesh on these theoretical bones, from the employment of traditional ethnographic methods (Revill 2004) to more novel methodological instruments as photographs and the daily diary (Latham 2003). Yet by and large "studies" employing Nonrepresentational Theory involve far more theory than research.

Feminist geographers, in contrast, focus on those embodied objects and subjects missed by Nonrepresentational Theory. Hayes-Conroy and Hayes-Conroy (2008: 466) put it as follows: "Neither Foucauldian poststructuralism nor nonrepresentational theory alone have been able to develop new understandings of immaterial and ubiquitous powers/forces (e.g., the disciplined body or the affective body) without abandoning, more often than not, the capacity to be politically decisive". Whereas the insignificant is the focus of Nonrepresentational Theory, the significant—selfhood—appears to most interest feminist geographers. That is because only the significant, it is believed, is political. Even Elspeth Probyn (1999: 221; see also 2001: 132), a well known feminist geographer who has written explicitly on the subjects of food and eating, is staunchly committed to elevating the construction of self above all else: "'Eating the other' is both a metaphor for imperial violence, and the point where knowing the self and caring for the self emerge, where food and sex emerge". The problem with this approach is that it is not radical enough. While I've found personal experience and biography too consequential to ignore (in this respect I am in agreement with feminist geographers) one cannot deny the power of the mundane.

I realize that I have yet to give a concise definition as to what "lived experience" means in the context of this book. At its most general level, the lived experience seeks to *enliven* social theory by injecting living, breathing, feeling bodies into social methods and conceptual frameworks (see Carolan 2009a). Even among those approaches in social theory that claim to be explicitly about bodies—like Nonrepresentational Theory, Actor Network Theory (ANT), and certain strands of feminist theory, etc.—one is often hard-pressed to hear from *actual* bodies, in terms of how they think and feel. On the other hand, embodied realism (*a la* Merleau-Ponty) is a little too focused on the (essentialized) body, ignoring those important relationalities that give shape to the lived experience. The lived experience, then, carves out analytic space for this lived (relational) experience without essentializing those experiences (more on this later), arguing that these experiences are sociologically and politically meaningful. This separates the lived experience from Nonrepresentational Theory, feminist scholarship, Foucaultian scholarship, and ANT. The lived experience seeks to give the reader a feel, literally, for *specific* embodied understandings, whereas the approaches mentioned previously prefer to speak of *general* embodiments, which can get them into trouble.

To think of it another way: the concept of the lived experience refers to a subject-decentered approach. This draws upon, but does not perfectly parallel, what is known among social theorists—such as Bruno Latour, Donna Haraway, and John Law—as a decentered approach. As Law (2002: 91) explains: "[A decentered approach] proposes that *objects are an effect of stable arrays or networks of relations*. The suggestion is that objects hold together so long as those relations also hold together and do not change their shape" (emphasis in original). But I think it would be more accurate to refer to this tradition as representing an object-decentered approach; the attention, after all, is on the production of *objects* (or what I refer to in another work as the production of "object-ivitiy" [Carolan

2010a]). A decentered approach ultimately places analytic—and for that matter ontological—primacy on becoming rather than on being. There are multiple ways, however, to enter into that relationality. The analyst can enter by detailing how those networks of relations (to use Law's words) hold together objective states. And there exists a diverse array of case studies that trace out the making of objects, from nuclear missiles (Mort 2002) to biotechnology (Carolan 2010a), water pumps (De Laet and Mol 2000), and airplanes (Law 2002).

Yet for all we gain by adopting this approach it leaves the "objects" from within—our thoughts, feelings, knowledges, and understandings—untouched. This is not an indictment against object-decentered approaches; I recognize no approach should be expected to do everything. It does, however, provide justification for the approach taken in this book. Just like one cannot divorce food from the relations that hold the objects of the food system together, it is equally a mistake to expunge those relations from the lived experiences that create stability and change in the network.

Chapter Overview

Chapter 2 works to set the stage conceptually as well as historically. In terms of the former, this chapter seeks to further familiarize the reader with what it means to think like a body. The myth of objective, disembodied, rational thought runs deep in Western thinking (see Whitehead 1967 for a thorough account of this history). Chapter 2 gives the reader time to adjust to this alternative understanding of knowledge before proceeding on to the case studies. It accomplishes this through two movements. In the first a brief reprieve from food is taken as I detail, historically, philosophically, and sociologically, how we got to this point, where we have to be reminded that we think as a body. The second movement brings us back to food, looking at how our embodied relations with food (at least in the West) have changed over the last century. I take time to discuss examples of how we have become tuned, collectively, to Global Food.

Chapters 3 through 5 hold the case studies. Customers and workers of two separate CSAs, visitors and workers of a heritage seed bank, and backyard chicken coop owners represent the empirical subjects of Chapters 3, 4, and 5, respectively. The diversity of cases across these three chapters occurred by design. Each offered participants a unique lived experience (though, as one would expect, overlap is inevitable) and thus each was found to challenge Global Food in unique and important ways.

These chapters are not interested in giving a descriptive overly-general definition about what these spaces are about. There are numerous books written on, for example, the history of CSA (see e.g., Henderson and Van En 2007) as well as on how to raise chickens in one's backyard (e.g., Damerow 2002). I have no interest in parroting that information here. What the reader will find in these chapters, rather, is a detailed discussion on how these *specific* spaces and the *specific* bodies therein embedded felt about phenomena related to food production and consumption. Note also how I do not conclude the previous sentence by saying "... *because of* these spaces and the embodied relations they make possible". The specific embodied knowledge attributed to these spaces is always tentative and suggestive, recognizing it is possible that certain understandings and knowledge could have preceded one's involvement in the space in question. In some instances, however, the epistemic significance of these embodiments appears fairly clear.

This brings me to Chapter 6. The focus of the book up until this point is largely on consumers, in terms of how we taste, perceive, and understand the food we eat. I could end the book here. Yet this would leave me unsatisfied were I the reader. After all, if consumers have become tuned to a certain way of *doing* food consumption could not the same be said about producers? Is not food production heavily dependent upon embodied knowledge? Beyond merely wanting to display some analytic balance by being able to say something about both "sides" of production and consumption lays a more practical concern. If we want to talk about tuning consumers to alternatives to Global Food we would do well to discuss ways to make sure there are individuals capable of producing food that are in line with these alternative tunings. Chapter 6 therefore makes an explicit turn to the production "side" of food. The aim of this chapter is to answer the question: how do CSAs, heritage seed banks, and backyard chicken coops serve as repositories for embodied knowledge to *do* food production?

Chapter 7, to make clear that the assemblages of the lived experience include broader structural phenomena, takes a closer look at the political economy of tuning, particularly as it applies to small-scale, local market oriented food systems. Our tuning toward Global Food was, and continues to be, made possible by broad socio-economic forces. Some of these connections are made in Chapter 2, where I attempt to socially, economically, and politically contextualize our shifting lived experience toward food over the course of the twentieth century. In Chapter 7 those networks are brought back into focus, noting that as we contemplate ways to elicit alternative lived experiences of food we might think about the role the state can and should play in this process.

Pushing the analysis outward in this manner also offers a bridge between an agro-food studies literature dominated by a political economy focus (see e.g., Weis 2007; Wright and Middendorf 2008) and food studies literature dominated by poststructuralism and more textual concerns (see e.g., Ferguson 2004; Johnston and Baumann 2010). This division has all too often stopped "analyses of the nature, culture and political economy of food [...] [from taking] place on the same page" (Freidberg 2003: 6). While this is not the only way over this impasse, I do wish to stress upon the reader that an analysis like this is not blind to the political economic realities that give shape to systems of food provisioning.

The following chapters were intentionally written to be read in any order—or no order at all if the reader is especially interested in the subject matter of one or two specific chapters. While there is a clear narrative thread that travels through all seven chapters, the reader will discover that each chapter can stand fairly well by itself and makes a distinct scholarly contribution in its own way. Thus, while the reader could read chapters selectively I encourage them to still read them all (though, admittedly, the order in which they do so is less important). I mention this now merely to prepare the reader for the somewhat unconventional journey that follows. Unlike the conventional narrative structure found in most academic books, where each chapter stands firmly on those that precede it (and where to skip a chapter or two can feel like going from arithmetic to differential equations), each chapter is, to some degree, self contained. Each chapter therefore begins by setting up, historically, philosophically, and sociologically, the phenomena that are later empirically unpacked with the aid of the case study. Doing this allows the book to serve multiple needs, depending on the interests (both theoretical and empirical) of the reader.

When taken as a whole, I hope to show that spaces like those discussed in the forthcoming chapters have "value" beyond that generally discussed by scholars. One popular framing of local food systems, for example, cites CSA, farmers markets, farm-to-school programs, and u-picks as creative market-based strategies that have the potential to strengthen the economic vitality of rural communities (see, for example, Bagdonis et al. 2009; Brown and Miller 2008; Hancharick and Kiernan 2008). Other research speaks of the resiliency that these artifacts add to a region's food provisioning capabilities (see, for example, King 2008; McCullum et al. 2005). That's all well and good. But might they also have value beyond what is recorded on an accountant's ledger or in a resiliency model? We can look to the marketplace and quantify the value of such things as CSA, heritage seed banks, and backyard chicken coops. Yet is that were their value ends? Is there anything about them that is non-quantifiably valuable? I can answer this question in the affirmative; though the reader will have to read on if they wish to learn more.

On Essentialism, Political Economy, and Representation

While the philosophy of substance argues that substance is distinct from its relations the stance adopted in this book rejects the argument that there is a fundamental rift between a thing's inner essence and its inconsequential surface fluxuations. What something *is*—its being—is an event of its relations. So to be clear: talking about taste and knowledge through the lens of lived experience does *not* essentialize these understandings, nor does lived experience represent the only way to talk about food politics. I also want it to be clear that this book does *not* seek to contradict or refute the agro-food studies literature, which, I contend, fails to think like a body. Rather, my intent is to compliment the political economy approach so popular among agro-food studies scholars. Agro-food analyses can't afford to ignore political economic forces. But the reverse, I insist, is also true: we cannot have a complete understanding of food politics without an understanding of how lived experience plays into knowledges, ethics, and feelings about food consumption and production (see also Goodman 2010a, 2010b). In the words of Latour (1993: 113): "nothing is, by itself, either reducible or irreducible to anything else. Never by itself, but always through the mediation of another". It is in this spirit that I investigate the lived experience: as just another piece to that great relational puzzle we simply know as "world".

I therefore agree with my colleagues who so convincingly argue that food is about class and political economies (see e.g., Allen and Kovach 2000; Coombes and Campbell 2002; Friedmann 2007; Guthman 2008a). Yet class, political economies, and structures do not a world make. Moreover, investigating how we think as bodies adds an important component to how we conceptualize (and ultimately enact) social change. Social change requires bodies not only endowed with resources, like social, economic, cultural, and political capital—an important take home message provided by a political economy approach. Social change also requires bodies that think social change *ought* to occur. How do bodies become tuned to the status quo (or to alternatives)? What makes bodies *want* change? These are important questions; yet they go largely unanswered in the agro-food studies literature. Lest we forget, at the heart of change/status quo are living bodies.

And it's *specific* embodiments that concern me here. Talk of the generalized embodiments that lurk within, say, Foucault's otherwise carefully argued works is avoided in this book as I seek to make sense of how respondents made sense of their food-world. Embodied knowledge is not something that is held by only specific bodies. We all think like a body—which is to say, relationally—regardless of our location in social space and time, ethnicity, and class. And if we all think relationally we all think about food relationally too. Ergo: if you eat you ought to be able to relate to the voices, arguments, and bodies conveyed through this book.

Finally, I realize some might take methodological issue with the emphasis I am placing on the lived experience, which, presumably, includes those aspects of life (e.g., sensations, knowledges, and feelings) that cannot be reduced to words. This begs the question: can social scientists even "catch" with their methodological nets this lived world and if so to what degree? How does one represent what is fundamentally un-representable? Or, as McCormack asks (2002: 470), "how, when such movement is often below the cognitive threshold of representational awareness that defines what is admitted into serious research, does one give a word to a movement [or sensation, experience, etc.] without seeking to represent it?" The world, and all of its "abundance" (Feyerabend 2000), can only be known partially. The representations—words, numbers, illustrations, and the like—we use cannot possibly contain it all. They inevitably will ooze. They will leak. But they still hold something. So I admit: we cannot literally feel what respondents experienced in their lived lives in the following pages. But this does not mean that we cannot at least get a taste of their world through their words. As Latham (2003: 2000) colorfully argues on the subject of employing traditional research methods to "get at" the buzzing world of the more-than-representational: "Pushed in the appropriate direction there is no reason why these methods cannot be made to dance a little". With that said, let's dance.

On Mushrooms (and Other Dichotomies)

I have a confession to make: I don't really think Global Food, as I call it, is as monolithic as my language makes it out to be. Note the organic-ification of industrial food—the organic Twinkie comes immediately to mind—or the conventionalization of organic agriculture—as evidenced by the explosion of organic food sales in big-box stores like Wal-Mart. Dichotomies like global/local, fast/slow, and healthy/unhealthy are not as self evident as the debate over food sometimes makes them out to be (a point elaborated upon in Chapters 6 and 7).

Having spent much of my career unpacking blackboxes—Global Food, for instance, is such a blackbox—I also realize the necessity of keeping some of those boxes "closed", at least partially, otherwise there would literally be no-*thing* to talk about (see Carolan 2010a: 5–6). Global Food, I realize, can be done in a more socially just and environmentally responsible way. Likewise, there is nothing inherent about more local forms of food production that make them "better" than those forms made possible thanks to longer commodity chains. The point of this book, however, is not to assign liability or blame to one mode of production over another. I'm interested, rather, in why I initially preferred canned mushrooms over fresh and why the reverse is now true. I could get bogged down trying to answer such thorny theoretical questions as "What *is* a fresh mushroom?" But I won't. Just like I won't get bogged down trying to define—in precise, positivist, materialist terms—what *is* Global Food? Besides, following the aforementioned (subject-/object-) decentered approach that underpins my view of the world, what matters are *relations*. So it's relations that I am interested in detailing; relations that bring forth certain competencies, knowledges, and feelings about what we eat. Recognizing that my "being with" canned mushrooms as a child shaped my understanding of this food I hypothesize that such relationalities underlie all (food) knowledges, making those connectivities inherently political. The following chapters test the validity of this hypothesis.

Lastly, this book contains no "if they only knew" arguments (Guthman 2008a); a naïve position found lurking in some food scholarship (see e.g., Waters 2008) that goes something like this: if people only possessed more explicit (representational) knowledge they would make "better" choices—namely, they would choose something other than Global Food. As I detail, it is not about knowing "more" or "less" but knowing *differently*; knowing *through* different relations. This book is about understanding what those relationships are, and how those relations came to be, that shape our food choices. It does not wish to cast judgment on individuals for relating to food as they do.

Chapter 2
Some Backstory

Merleau-Ponty (1992: 89) argued a half century ago that we have largely "unlearned" what it means to think like a body. This is not to say we don't think like a body, just that we've become used to seeing thought as a thing (rather than as relations). While I do not expect a chapter to undo this forgetting, I hope it will soften some of those grooves of the mind that keep us from grasping how embodiments matter, thereby preparing the reader for the sensual understandings that await in later chapters. After laying this additional conceptual groundwork the chapter returns to the subject of food. The chapter concludes by providing some historical context to the argument that changes to the food system in the twentieth century altered our relationship with and thus knowledge of what we eat.

The Sensuous in Social Thought: A Brief Overview

In *The Spell of the Sensuous* David Abram (1997: 102–23) does some interesting philosophical detective work. The mystery he sets out to solve involves a noticeable epistemological shift that occurred in Ancient Greece around the time of Plato. It was at this time that knowledge went from being understood as thoroughly embodied to something premised upon the existence of disembodied, rational faculty; a Cartesian split—whereby mind and body are viewed as being housed in separate worlds—almost two millennia before the namesake of this phenomenon was born. The Homeric poems (which date from well before the time of Plato) describe an earth that is alive, a world that is consequential, and an ontology that is smeared. In these epic stories, natural events and human emotions are as perceptibly indistinguishable as two clasped hands:

> an army's sense of relief is made palpable in the description of thick clouds dispersing from the land; Nestor's anguish is likened to the darkening of the sea before a gale; the inward release of Penelope's feelings on listening to news of her husband is described as the thawing of the high mountain snows by the warm spring winds, melting the frozen water into streams that cascade down the slopes (Abram 1997: 103).

The Homeric self has been described as indeterminate, which is to say it was not perceived as an object separate from the world (Brown 2006: 24). Yet by Plato's time this self became cemented; fixed into an unchanging, objective entity. Abram sets out to explain why this shift occurred. Or to put it in terms more befitting a mystery, he seeks to determine what killed the sensuous understanding of knowledge that persisted in Ancient Greece until the time of Plato.

Abram's prime suspect: the Greek alphabet. First invented (actually adapted from the Semitic *aleph-beth*) centuries before Plato, it was slow to become widely adopted. Pre-alphabetic Greeks preserved tradition, culture and knowledge orally. Theirs was a richly oral society. Those who helped tell these stories were called rhapsodes (or in modern usage, rhapsodist). The alphabet's slow diffusion was linked, as least in part, to a belief at the time that there was no need for this new "technology". Ceremonies allowed these living encyclopedias to narrate the past. Moreover, social status was ascribed to those with the skill to recite and perform. Ancient Greece was organized significantly around this oral tradition. And as we know from the field of science and technology studies, when society is organized around a particular technological form that technology becomes, to a degree, self-sustaining and resistant to change (see e.g., Carolan 2009b; Hughes 1969).

Rhapsodes did not memorize their stories. They did not have to. The poems were wrapped up in rhythms and bodily pulses and synced to breathing patterns, which made the "mental" act of memorizing unnecessary. The poems of this time had these qualities because, in the words of Abram (1997: 106), "pulsed phrases are much easier for the pulsing, breathing body to assimilate and later recall than the strictly prosaic statements that appear only after the advent of literary". Ever wonder why that old nursery rhyme you learned as a child is so easy to remember? But even our childhood nursery rhymes are written down, so they remained fixed across time and space. There is speculation that Homer's *Iliad* and the *Odyssey* were originally, in pre-alphabetic Greece, fluid stories that varied from rhapsode to rhapsode (Havelock 1986). Only after being written down did these poems come into existence as objective text.

The very idea of memorization may itself be a product of an alphabetic culture. To memorize something presupposes that there is some*thing* to memorize that is independent of the speaker. In pre-alphabetic Greece, the notion of ideas existing apart from a living body was, at least according to some scholars, inconceivable. As the famous twentieth-century classicist Eric Havelock (1986: 112) once explained:

> It is only when language is written down that it becomes possible to think about it. The acoustic medium, being incapable of visualization, did not achieve recognition as a phenomenon wholly separable from the person who used it. But in the alphabetized document the medium became objectified. There it was, reproduced perfectly in the alphabet [...] no longer just a function of "me" the speaker but a document with an independent existence.

The point of mentioning this is to emphasize the difference between ways of knowing that hinge upon representations (like words and pictures) and those that involve lived experience. Having bodies so "tuned" today to the visual, as some research argues (Kress and Leeuwen 1996: 33–40; Rose 2001: 6–7), it is admittedly difficult to put ourselves in the shoes of someone whose ways of knowing are more sensual. Perhaps this is why early European explorers thought the Inuit (the native peoples of northern Canada) defiant due to a perceived unwillingness to comply with requests to draw maps of their territory: because they assumed the Inuit were similarly tuned to know the world. Unlike their Western visitors, the Inuit privilege input from the ear and thus "define space more by sound than sight" (Carpenter 1973: 33). Rather than an act of disobedience, the Inuit simply did not think about the world in a way that allowed them to produce a two-dimensional map. Rodaway (1994: 110) explains why the visual in such a space as the Arctic has less informational value:

> The wind was perhaps more important than the vista, offering environmental information from its noise, force and direction, and from its olfactory content as well. The long periods of darkness in the tundra winter and the snow and ice expanses where sky and land and sea merge make visual sensitivity less useful, especially when the individual is hidden well into his or her parka to keep out of the cold and biting wind. Instead, the other senses take on a greater importance; including the hearing of distant and invisible sound sources—the water against a shoreline hidden by fog, a subtle change in the sound of ice over which the sledge is moving, the tone of the wind as it howls. It is a land not of objects—that is the world of sight—but one of events and relationships—that is the all-round alertness of the ears.

Rather than coming to us as isolated bits of information, sound is integrative (Feld 1984: 163–5). The world is a symphony, full of consonance, dissonance, and harmonies, which we cannot easily disentangle ourselves from (as anyone can attest to who has "heard" a loud clap of thunder in their chest). Vision, conversely, is more prone to epistemic fragmentation, particularly when stripped of other sensorial information; a way of knowing that involves extracting phenomena from a background creating the appearance of isolation and independence (Jay 1994: 211–20). Within the Arctic Circle vision is of little help because there is either all background or all foreground, depending upon how one looks at it. Living in a space not oriented to the visual, the Inuit long ago adapted to these epistemic limitations.

Now think about how the industrialization of food tuned our bodies to know food differently from the grip obtained when food production was known—literally—firsthand. Knowledge that can be conveyed with words, pictures or a diagram certainly has its place. One of its virtues is that it travels well. Fixing information in space and time, either through pictures, words or numbers, allows others to know the same thing. Representational knowledge, at least according to some scholars, lies at the heart of conventional agriculture with its emphasis upon highly transportable knowledge, such as feed charts for livestock and standardized NPK (nitrogen, phosphorous, potassium) application rates (see, for example, Cohen 2009; Kloppenburg 1991; Kloppenburg and Burrows 1996; McAfee 2003). Similarly, the distance separating most individuals from the process of agro-food production leads consumers to rely upon proxies as they seek to understand what they're eating. Early food packages, some of which are discussed shortly, acted as proxies for proximity in that they attempted to convey information that could no longer be obtained firsthand by consumers about their food. Yet as proxies these representations are impoverished substitutes for the thing itself being replaced: the lived experience of food production.

Having grown up in the presence of gardens and asparagus beds, spending many afternoons engaging in such activities as weeding corn rows, picking green beans, and making coleslaw, I have acquired certain understandings—understandings that are not better, just different—that many of my friends lack. A few months back I ran into one such friend at a grocery store. In their cart I saw a crookneck summer squash that was, to me, past its prime. For one it was too large. And after holding it in my hands for a couple seconds I also learned that it was beginning to feel spongy. I asked my friend if this was the only one left (it must be, why else buy it, I thought to myself). No, they told me. After exchanging a few words about the squash, and vegetables in general, I learned that my friend had a different understanding of how to select "good" summer squash. He believed, for instance, that those available in the store were, in his words, "all the same". He also told me how he usually picks the biggest one available so that "instead of cutting two there's only one big one to prepare". At one point in our conversation he joked how he wished there was a "sell by" date on fresh vegetable so he would have some way to determine their freshness. When we went our separate ways I began to process what had just transpired. I was struck by my friend's apparent definition of "fresh"—at least as it pertained to crookneck summer squash—thinking how such a definition works so clearly in favor of the conventional food system. Juxtaposed to my friend's definition, I then began to think about how my understanding of freshness made it difficult for me to find a suitable summer squash at the grocery store. In fact, I almost exclusively get them from either my garden or local growers. Almost all others look, feel, and taste "unsatisfactory" to me.

Some historians postulate that Western thought underwent a profound shift in the kinds of ideas available to the human mind at the point Greek philosophy shifted from oral to literature form (see e.g., Havelock 1986; Ong 2004). Though I am not suggesting anything as profound or worldview-shifting as this, I do think the industrialization of agriculture produced a shift in our knowing of food by fundamentally altering our relation to food production. And this alternation, in turn, has served to perpetuate the system that gives life to this understanding.

The Sociological "Trap" of Taste

Western philosophy long ago established a hierarchy of the senses. Assumed within this hierarchy is the belief that distance—namely between knower and known— has a cognitive, moral and aesthetic advantage, while a bodily sense like taste brings one dangerously close to the object of perception (Korsmeyer 1999: 12). Plato, in his famous discussion of the cave in *Republic* VII, uses a wealth of metaphors to convey the intellectual power of sight: shadows, light, the sun, the darkness of a cave, and Forms. Descartes too, particularly in *Meditations*, famously extolled the virtues of sight. The belief, then as now, has been that vision, though open to creating illusions, is freer from the pull of emotions and appetite than other more corporeal forms of knowledge, like taste. Sight fosters distance, or so it would seem. It separates mind from body, self from world, giving the illusion of a literally autonomous rational facility capable of exploring not only the world around us but also more divine regions of the universe were bodies cannot travel. For Plato, the philosophic life required the denial of these lower senses. In his *Symposium*, for example, friends gathered for a banquet. But as they philosophize about love they made sure not to eat or drink so as to keep their intellect sharp. And similarly, the denial of the more fleshy lower senses has been a constant theme in theological scholarship for over two millennia (though some texts from the Middle Ages point to a brief exultation of taste as the true way to know the nature of things [Burnett 1991]).

It was not until the twentieth century that scholars began to view taste as more than a form of knowledge rooted in a pre-social body driven by impulses and desires. Social scientists were first to write on how taste is itself a cause and consequence of socio-material organizational patterns. Norbert Elias (1996, 2000) devoted most of his career to uncovering the sociological underpinnings of taste. Elias evoked the term "habitus"—well before Bourdieu popularized its use—to speak of the internalization of collective (social) dispositions, feelings, moods, and manners. In their Preface to the book *The Germans*, Eric Dunning and Stephen Mennell (in Elias 1998: ix) note how habitus, as used by Elias, could be explained as "embodied social learning". For Elias, our feelings about what constitutes good and bad manners, proper and improper behaviors, and civil and uncivil tastes and dispositions are ultimately social in nature. Yet their social nature does not make them any less "real". Over time, through repetitive practices, the result of which gives these social constructions the force of objectivity, the origins of one's habitus become less transparent. Elias provides considerable insight into this process, specifically in regards to those feelings and tastes that pertain to manners and our understandings of civility.

For Elias, tastes do not occur randomly. Taste is strategic. According to Elias, these dispositions have been used to maintain, and when possible expand, class divisions in society. To put Elias' argument as simply as possible: socially desirable tastes often take time, and perhaps even money (e.g., etiquette lessons), to acquire. This point has also been made by the social theorist Pierre Bourdieu (1984: 13; 1995: 187), who spent much of his career talking about the strategic role of taste. For Bourdieu, taste is like a type of capital—what he called "cultural capital". Bourdieu, though he differs in some of his details, makes many of the same points as Elias. For both these theorists, taste orders society; it organizes people into groups and helps define group boundaries and individual identities.

Elias and Bourdieu were interested in the process of social reproduction. Taste, as both theorists powerfully detail, helps structure society, giving it the appearance of stability over time. This is not to argue, however, that taste is immutable. Nor is taste to be equated only with social statics. Taste also has emancipatory powers.

I therefore talk about "taste" fully cognizant of the sociological pitfalls that accompany the concept, especially as it is applied to food. Too often taste is evoked by such authors as Michael Pollan and Alice Waters—two bestselling food writers—in a very essentialistic (and sometimes condescending) way. The subtext of their argument goes something like this: if people only knew what an organic tomato tastes like or how conventional food is grown or the environmental footprint of that fast-food cheeseburger or … —if they only knew they would adopt the "correct" eating habits (Guthman 2008a). This is *not* what is being argued in this book. Discussing taste and knowledge through the lens of the lived experience need not essentialize these understandings. Moreover, understanding food through the lived experience fundamentally rejects the information deficit model tacitly assumed by writers like Pollan and Waters. Supplying individuals with explicit (representational) knowledge about what they are eating will only minimally impact their feelings and understandings toward food. If the goal is to decentralize the food system and create shorter agro-food chains then we'll need to ask what lived experiences would make Global Food out of tune. We need to make people *want* an alternative. Simply knowing about the problems associated with the current system is not enough to change behavior and elicit collective mobilization.

Michael Polanyi has written explicitly on the limits of representational knowledge with his discussion on the tacit dimension. Polanyi (1966: 4) famously pointed out that "we know more than we can tell" in his discussion of tacit knowledge—that component of knowledge that cannot be reduced to words, numbers, and/or figures (what Polanyi called explicit knowledge). One of the most compelling examples Polanyi offers of tacit knowledge is riding a bike. Even the most accomplished bicycle rider will have a difficult time converting to words their knowledge on the subject. Polanyi is not, however, claiming that there are two independent types of knowledge, one tacit and one explicit. Rather, what he is saying is that *all* knowledge has a component that cannot be codified—what he calls the tacit *dimension* of knowledge. In Polanyi's (1966: 20) own words:

> The declared aim of modern science is to establish a strictly detached object […] [however if] tacit thought forms an indispensable part of all that knowledge, then the ideal of eliminating all personal elements of knowledge would, in effect, aim at the destruction of knowledge.

What this teaches us is that *practice* is inseparable from knowing. Perhaps not surprisingly, given its interest in innovation and knowledge transfer, the business management and organizational science literature has found the phenomenon of tacit knowledge to be of particular importance. Studies have since documented the significance of this dimension of knowledge in a host of practices: among, for example, butchers and midwives (Lave and Wenger 1991), photocopy repair technicians (Orr 1996), and engineers charged with designing and building a plant for the chemical compound phthalic anhydride (Spitz 1988). As two leading organizational scholars argue, the importance of the tacit dimension for the diffusion and innovation of *all* types of knowledge lies in the fact that "only by first spreading the practice in relation to which the explicit makes sense is the circulation of explicit knowledge worthwhile" (Brown and Duguid 2001: 204). This is one reason why it is important to focus on the placement of bodies in social space and time: because such a focus not only reminds us of ways of knowing that are often forgotten (precisely because they cannot be easily codified) but it also enriches our understanding of all knowledge. Making space for this dimension of reality enriches our sociological imagination. Looking at the relationship *itself* as a living, fluid, non essential social fact broadens the analyst's grasp of the world by giving them the means to understand how understanding, knowledge, and emotion often originate from *outside* the body, from the relationship space itself; a "setting [that] is cancelled out by such methods as questionnaires and other such instruments" (Thrift 2004: 60).

I am well aware that I am going against a long lineage of thinkers who have been steadfastly suspicious of things like taste, emotion, and affect. I recognize that there are dangers in talking about taste, the ephemeral, and the visceral. But we face equal dangers when we don't. One seminal work in what has come to be known as sensuous scholarship is Stoller's (1989) analysis of the Songhay of Niger. Disheartened by the bland style of conventional ethnographic research and by his own initial inabilities to grasp the sensuous dimension as a scientist, Stoller (1989: 9) felt the need to write "ethnographies that combine the strengths of science with the rewards of the humanities". According to Stoller (1989: 9), it's the relationality of the experience—an experience full of sounds, smells, tastes, textures, sights, and affects—that makes sensuous scholarship radically empirical and thus renders "accounts of others more faithful to the realities of the field—accounts which will then be more, rather than less scientific". I also feel strongly that research *about* the senses must also be *for* the senses and avoid the dullness of overly formal, overly analytic, and overly anonymous scholarship. This, I am sure, will make well trained social science "technicians"—as C. Wright Mills (2000: 212) pejoratively calls them—uncomfortable; who, though well trained, are only trained in "what is already known". After all, most social scientists (myself included) are trained to think and write analytically and less so evocatively and imaginatively (Mills 2000: 212; Vannini et al. 2010: 380–82). Rather than try to undo that training with a few thoughtful sentences I will let the following analysis do the convincing for me. By the book's end I am confident the reader will see (and feel) the value of social scientific scholarship that looks to the lived experience for understanding.

Why I'm Not a Romantic

It is commonly assumed that we think *about* things, never *through* things. But this view clings too stubbornly to the Western assumptions that locate mind separate from body; the view of a disembodied, rational faculty that dates, as earlier mentioned, to Ancient Greece. Sherry Turkle (2007, 2008a, 2008b, 2009) makes a compelling case in a four volume edited collection of essays on how we think through things and on how objects make us who we are. Along similar lines, Don Ihde (2000) has shown how machines help us think theoretically. Ihde coined the term "epistemology engine" to remind us of this cognitive attachment that we have to the things around us. Using the example of the steam engine, Ihde explains:

> [T]echnological innovation inaugurated revolutionary changes in theoretical understanding. It paved the way for advances in calorific theory that in turn led to the development of thermodynamics. In this sense: The machine, not raw nature, suggested the phenomena. In sum, the concept of 'epistemology engine' appears to be a theoretical extension of the phenomenological insight that practical coping tends to precede theoretical reflection (Ihde and Selinger 2004: 363).

Notice how this smeared approach to cognition also places into question the very category of "technology". Examples of these blurred ontological lines between humanity and technology are everywhere. Tools, the structures of bones and muscles in the hand, and neural systems are believed to have evolved in lock step fashion, producing a self reinforcing loop. Tools led to greater manual dexterity and the formation of neural pathways to accommodate their use, all of which, in turn, lead to more complex tool usage, and so forth (Thrift 2005: 70–74). Phenomenologist Merleau-Ponty (1992: 143) also provides a number of excellent examples of this indeterminacy. The most notable involves the blind person and their cane, and how through repeated use the cane becomes absorbed into their living body. As Merleau-Ponty explains, when probing the ground, the unpracticed person only feels the ground through the cane. As habituations slowly lead to the absorption of the cane into the body, however, more sophisticated forms of interaction develop. One begins to feel the ground directly, in which case the stick ceases to be an object in itself and becomes instead part of the body. The integration of the objective world into the body of subjective experience challenges beliefs that "the body" ends with the flesh.

The implications of this empirical insight run deep. For if technology is part of humanity, rather than apart from it, than there is no going back. Thoreau's noble savage, as it turns out, was no more removed from the technological world than we are today. Ihde eschews what he calls the "Edenic" approach to technology. Among some of his examples of how technologies have *increased* our lived experience: the oven—a type of externalized stomach that has introduced to our bodies an array of lived experiences (such as novel odors, tastes, textures, and sights). Or take the speed boat:

> The modern high-technology boat, precisely in its capacity to allow one-self to be embodied through it, place one more closely in turn with wind and water

than was so through the insulated and dampened result in the resistance-to-maneuvering of the older wooden vessel (Ihde 1990: 164).

But this is not an argument for technological optimism either. While a lived approach does not allow for the drawing of clean lines between a body and its technologies it still allows for distinctions to be made between the types of sensitivities that these relations bring forth. While technologies may indeed amplify our senses, this sensitivity comes at an expense. A speed boat may heighten our relation to wind and water but it reduces our ability to see the landscape nearest to us. Using the internet, I can "travel" to distant lands with a few clicks of the mouse. But the smells, sounds, and feelings I experience while "visiting" remain those of my office. And with binoculars, we literally cannot see the forest for the trees.

I am fully aware that we can never "go back". And the same holds for how we *do* food consumption and production. To recreate past food embodiments is impossible, unless the entire world was somehow shaped in the past's image. For example, the very fact that food was not thought of in life-or-death terms for any of the respondents in the following chapters demonstrates a fundamental shift in lived experience from previous generations. I recently read *O Pioneers!* and *My Ántonia* by the early twentieth-century writer Willa Cather. These stories are set in rural Nebraska in the late 1800s. Thinking about food as I do, I was struck by how the books subtly chronicle the ever-present fragility of corporeal nutrition for this previous generation. Cather's characters frequently engaged in activities involving, in some way, food. The books speak repeatedly of the centrality of food scarcity and security among immigrant farm families as they settled the Nebraska prairie. Many summer and fall days were spent gathering, foraging, and harvesting all types of edible artifacts in preparation for the long, harsh winters of the plains. Those ill prepared risked starvation. I cannot help but think that this embodied reality shaped our ancestors' understandings of food. Those interviewed for this book, conversely, knew not this fragility. They knew that if their chickens died, if their corn did not germinate, or if their CSA went out of business, they would still eat.

Just as one cannot step outside of socio-material relations to understand the mountains of Colorado, there is no such thing as a pure experience of food. Previous understandings of food were no less mediated than those of today. Thoreau's romantic savage is a myth. The empirical cases examined in the following chapters thus do not represent a stepping back or a returning to. What they do offer, and this is an important point, are (embodied) experiences that threaten the dominant system of food production and provisioning. They each are sources of a [grip] that make problematic the sensations associated with Global Food. Or, to put it another way, they each represent spaces through which Global Food loses a bit of its grip over us.

"Tuned" Bodies are Made (Not Born)

Siderodromophobia (derived from modern Greek "siderodromos", railroad, and "phobos", fear): "a more or less intense spinal irritation, coupled with a hysterical condition and a morbid disinclination for work, which, as a result of shock, occurs among railroad employees, who, in consequence of their occupation, are especially predisposed to it" (Medical and Surgical Reporter 1879: 544). Soon after people began traveling by train reports emerged of individuals getting ill from the experience. More commonly known as "train sickness", the symptoms included, but where not limited to, eye infections, miscarriages, hemorrhages, vomiting, loss of vision, and an array of mental maladies (Kockelkoren 2005: 148–9). Some of these cases sound a lot like what we would call motion sickness today. Early train travelers were some of the first people to experience the somatic sensations that accompany being enveloped in a capsule hurtling through space. And just like today, these motions made people ill. Yet I cannot help but wonder if there was not more to this "illness" than an imbalance of inner ear fluid.

In a letter to his daughter dated August 22, 1837, poet Victor Hugo described his first experience of traveling by train:

> The flowers by the side of the road are no longer flowers but flecks, or rather streaks, of red or white; there are no longer any points, everything becomes a streak; the fields of grain are great shocks of yellow hair; fields of alfalfa, long green tresses; the towns the steeples, and the trees perform a crazy mingling dance on the horizon; from time to time, a shadow, a shape, a specter appears and disappears with lightning speed behind the window: it is a railway guard (quoted in Weiss 1998: 90).

Before people began stepping onto trains for the first time in the 1800s they had always experienced travel as part of the landscape rather than a part from it. On a train, people began to experience travel as passing *by* even though they were physically passing *through*. This experience must have been surreal, where sights no longer accompanied sounds or scents (save for those emanating from within the train) and where objects farthest away now became the easiest to see. Is it not at least plausible that some of the malaise associated with siderodromophobia originated from this lived detachment that train travel introduced into the world?

"Porridge or Prunes, Sir?" So begins an essay by the early twentieth-century English novelist E.M. Forster. In this essay, Forster describes a trip home to England via railcar. Forster is sitting in the Restaurant Car; a new import from the United States. Forster tells of his disgust when asked by the restaurant attendant those aforementioned four words:

> Porridge fills the Englishman up, prunes clear him out, so their functions are opposed. But their spirit is the same: they eschew pleasure and consider delicacy immoral. [...] That morning they looked as like one another as they could.

> Everything was grey. The porridge was in pallid grey lumps, the prunes swam in
> grey juice like the wrinkled skulls of old men, grey mist pressed against the grey
> windows (Forster 2002: 405).

This was followed by fish (which "was covered with a sort of hard yellow oilskin, as if it had been out in a lifeboat"), sausages and bacon ("[t]hey, too, had been up all night"), and toast (which was "like steel"). Finally came the bill, which, as Forster describes, "I paid dumbly, wondering again why such things have to be [...]. We often eat well in our homes or in clubs or in small restaurants which have not yet been spoiled [...] [by being made into] absolute muck".

One could easily read this essay as an indictment against industrial food. Yet my interest in this piece lies elsewhere. There is a body lurking in this piece: Forster's. And it is not tuned (yet?) to the food embodiments that will become ubiquitous by the century's end. In the words of Forster's body (and stomach):

> Breakfast could not be served until the train started. [...] I opened my book and
> tried to read [...] [b]ut I could not attend to the exquisite prose; the fever, the
> loveliness, the tenderness in hatred, the light and the scents of the south, would
> none of them come through. Breakfast, oh for breakfast! [...] At last the engine
> gave a jerk, the knives and forks slid sideways and sang against one another
> sadly, the cups said 'cheap, cheap' to the saucers [...] [and] the attendants came
> in crying 'Porridge or Prunes, sir?' Porridge or Prunes? Breakfast had begun.
> That cry still rings in my memory. It is an epitome—not, indeed, of English
> food, but of the forces which drag it into the dirt (Forster 2002: 405).

The lived experience described by Forster is a metaphor for broader social changes occurring at the time the essay was written (it was first published in 1944). Food is not a thing; its being is constituted through relations. Had those grey prunes and lumps of porridge been experienced by Forster within a familiar home I wonder if they would have evoked the same level of visceral disgust. Note that Forster spends as much time bemoaning the eating-in-the-railcar experience as he does expressing disgust over the food itself. As Sidney Mintz (2002: 29, emphasis in original) argues, "what a Whopper *is* is not forecast by its intrinsic nature but by the cultural matrix within which it is perceived and consumed". I feel comfortable arguing that Forster's abhorrence stems not from the food *itself*. The actual molecules that together made grey-looking prunes and porridge were not the actual source of his angst. The abhorrence, rather, lies in the lived experience of eating on a train in England in the twentieth century and in how this lived experience is haunting an increasing number of his eating haunts. The fear for Forster is that his (eating) world is becoming a Restaurant Car and the Restaurant Car the world.

Scholars have already noted the relationship between food and practices of "dwelling-in-traveling" and "traveling-in-dwelling" (Gibson 2007: 5) by illustrating how mobilities shape the lived experience of eating. The average American now eats one out of every five meals in his or her car (Trubek

2008: xiii). Certainly this must be of some consequence to how these bodies know food. This question is, admittedly, an empirical one and reappears in the next chapter. Yet even in an unanswered state it begs us to reflect upon how industrial food, like train travel, has introduced the world to certain competencies. And in both cases, these experiences now have a taken-for-grantedness that makes it easy to think they are natural and objectively self-evident. In reality, our understandings of phenomena like "food" and "travel" have adjusted with our embodiments. Which begs the question: have these adjustments occurred at the expense of certain knowledges lost? Which, in turn, begs additional questions: precisely *how* has our relationship with food changed and *who/what* are responsible for these changes? The remainder of the chapter, while speculative (Rosenthal 1986), speaks to these questions.

The Changing "Face" of Food

> "This look recent to you?" he said. He held the meat out toward her.
>
> "It looks bloody", she said.
>
> "I can't tell if it smells good. They got it wrapped up in all this goddamn plastic. You couldn't tell the working end of a skunk with this stuff on it".
>
> "I didn't know you ate skunks".
>
> "That's what I'm talking about. I can't tell what I'm eating with this goddamn plastic wrapper around it. It ain't like our own beef from the meat locker—when we get it I know what I'm getting". He shoved the pork roast back into the meat case and picked up another package. He held it close to his face, sniffing at it, grimacing, his eyes squinted. He turned it over and peered suspiciously at the underside.

This fictitious exchange comes from the award-winning novel *Plainsong* (Haruf 1999: 161). As a Generation Xer (born in 1974) I never knew a time when meat didn't come wrapped in plastic or when vegetables and fruit (and of course mushrooms) were not packaged in cans. For me, there has never been anything odd about walking by the meat case and being unable to smell its contents. Nor did I find it suspicious that the only sensorial information to come from canned goods was from the wrapper enveloping the can and the sound and feel of the contents swashing around as I shook the cylinder in my hands. But that is a product of my lived experience, which is premised upon the take-for-grantedness of conventional food. Had those lived experiences been like the gentleman's at the beginning of this sub-section, I would likely have similarly distressing memories about those food items purchased for the first time at a grocery store.

The story of the changing structure of agriculture, and of the growing distance between producers and consumers (Duffy et al. 2005; Eden et al. 2008; Princen 1997), is likely familiar to most readers. Today, less than 1 percent of the population in the US claim farming as an occupation, while about 2 percent of the population actually live on farms. The farming population in 1920, when the official Census data collection began, constituted 30.2 percent of the total US population. From then to now, the number of farms has dropped, while the average size of farms has increased. I would like to avoid discussing the various reasons *why* such changes have occurred. I do this not because I fail to find the subject interesting. But the story of the changing structure of agriculture has already been told (and retold) from a variety of different angles (for some classic analyses see Buttel, Larson, and Gillespie 1990; Cochrane 1993; Strange 1988).

Lost sometimes in this discussion of what is known commonly as the industrialization of agriculture are the equally momentous technological "advances" in food preservation. Indeed, were it not for the ability to feed people "freed" from agriculture who went to work in the cities there would have been little need for the increases in agricultural productivity that were witnessed throughout the twentieth century. Thanks to food preservation techniques, people could be separated by great distances from the land, farms, and farmers that fed them. Arguably the first modern food preserving innovation (recognizing that people had been salting, smoking, fermenting, and drying food for millennia) occurred in 1809, when Nicolas Appert won 12,000 francs from Napoleon for inventing a way to preserve food in bottles (a technique that Napoleon immediately put to use to feed his troops on the battlefield). In 1819, William Underwood put the same principles to use in the US but used instead tin canisters—hence the term "canned food" (Blay-Palmer 2008: 18). Add to that Louis Pasteur's pasteurization process invented in 1862, and the pieces were in place for a food revolution. Indeed, according to historical accounts the shift over to industrial foods was quick:

> The cook's store cupboard and larder in the middle class city home of the 1890s was as full of packets and cans as any housewife's now. With the help of bottled sauces, canned vegetables and fruit, and essences, she could choose between as many flavours (if less subtle ones) as a skilled chef who still made all his kitchen "basics" by hand (Renfrew et al. 1985: 10).

It is no secret that people initially viewed industrialized food with a degree of suspicion. Previous to this, people only knew food as an artifact produced either by themselves or by someone they knew. Understandings of freshness and quality were consequently wrapped up in a particular set of corporeal relations, which usually involved people being in intimate contact with what they ate. So, understandably, when food began coming in a can or box most people initially lacked the embodied resources to deal with these novel artifacts. For example, W.K. Kellogg believed early on that consumers needed symbols they could identify with that would give them confidence about the quality of the food they were buying. He began in

1906, therefore, to print his signature on every box of corn flakes in an attempt to build trust in the Kellogg brand (Blay-Palmer 2008: 46). In 1907, Kellogg placed an actual face on their product. Called the "Sweetheart of Corn", the illustrated image was of a young girl (looks to be between 10 and 15) holding an ear of corn in each hand. Between 1905 and 1940, Kellogg spent $100 million on advertising in an attempt to further assure consumers of the quality of their product (and of course to build brand recognition). The industry as a whole followed suit and by 1940 Americans consumed nearly as much processed foods as fresh food (Blay-Palmer 2008: 48).

Given the stance of medical and health professionals today towards processed foods it might surprise some to learn that this early push towards industrial food was helped along by these very professionals at the turn of last century. The argument, made by such industrial food pioneers as W.K. Kellogg and C.W. Post, was that the early pioneer diet was too rich for the sedentary, urban lifestyle (Blay-Palmer 2008: 45). What the "modern" body needed, or so it was argued, was access to standardized, safe, and healthy processed foods. This argument was used quite effectively and led, for example, to the flaked cereal revolution at the beginning of the twentieth century (Powell 1956: 85–6).

Though impossible to interview individuals from the nineteenth century about how they knew food, cookbooks from this time period provide a lasting proxy for at least some of this knowledge. For example, cookbooks from the late 1800s reveal a very different understanding of "fresh", and the practices for assessing this, than most today have in mind (and body) as they walk through the aisles of the local grocery store. Describing how to select fresh fish, one cookbook writes: "When fresh, the eyes of fish are full and bright, and the gills a fine clear red, the body stiff and the smell not unpleasant" (Wecox 1885: 506). Another cookbook offers the reader the following detailed account of how to determine freshness for a variety of fowl:

> In choosing poultry, select those that are fresh and fat, and the surest way to determine whether they are young, is to try the skin under the leg or wing. If it is easily broken, it is young; or, turn the wing backwards, if the joint yields readily, it is tender. When poultry is young the skin is thin and tender, the legs smooth, the feet moist and limber, and the eyes full and bright. […] Young ducks and geese are plump, with light, semi-transparent fat, soft breast-bone, tender flesh, leg-joints which will break by the weight of the bird, fresh-colored and brittle beaks, and wind-pipes that break when pressed between the thumb and forefinger (Gillette 1889: 70).

Note the level of full sensorial engagement involved: smelling, pinching things between the thumb and forefinger, and manipulating to see the whole animal—not "cuts" or "parts" as is typically the case today—from all angles.

Companies knew early on the importance of savvy marketing for gaining consumer trust. Consumers, for their part, sought out this knowledge too. Looking to fill in these epistemic gaps brought on by an increasingly industrialized food system, consumers turned to labels and advertisements to make sense of food. Early food manufacturers constructed a "founder" and a "founding story" through which they communicated their product. With the passage of the 1870 federal trademarks law, firms could protect images and narratives designed to provide a proxy of sorts for the lived experience that previously constituted peoples' grip on the food system. The Quaker Oats Man was "born" in 1877, followed by Aunt Jemima (1905), the (earlier-mentioned) Sweetheart of Corn (1907), the Morton Salt Girl (1911), Betty Crocker (1921), and the Jolly Green Giant (1926). All sought to place a face on food that had been removed by the industrialization process. In reality, these images and narratives presented little more than idealized sentimentalities. Illustrations of cows grazing in an idyllic pasture graced containers of evaporated milk and butter. Pictures of grain fields dotted with quail could be found on cereal boxes. And free-range pigs with a white barn in the background were depicted on some ham and bacon containers.

Most early food packages also contained text—and by today's standards a surprising amount of it—to give the consumer further "facts" about the food they were thinking about purchasing. In some cases, this text sought to convey the very sensorial information that the industrializing food system was denying. For example, a Sunkist advertisement from a 1917 issue of *Good Housekeeping* describes how their oranges are "luscious, tender and heavy with glistening juice" (see Figure 2.1).

From the perspective of a body living in a world saturated with symbols and the five second sound-bite, I have a difficult time understanding why firms a century ago would have bothered offering so much detail within their ads. Although they must have been effective, otherwise this text-saturated strategy would not have been so widely used, I doubt many consumers today would take the time to read a scrip the length of an entire page to learn about, say, a particular brand of honey (see Figure 2.2). Perhaps this gives some insight into the level of suspicion people at the time had of food raised without a face; so much that they actually took the time to read the stories contained in some ads in the hope of learning about their food.

Seeing some of these images and reading the accompanying text I often question the accuracy of these narratives. Nevertheless, even if firms attempted to accurately represent the farm-to-fork story of their food, certain knowledges were always missed. That is the uncomfortable truth about lived experience: the embodied knowledge it produces cannot be wholly reduced to, and thus cannot be reproduced by, words, pictures, a catchy jingle, or a figure point.

Food has always come with a story, which in part explains its power to produce and reproduce culture and identity (Allen 2004; Freidberg 2003). In the past, and in the case studies discussed in the following chapters, these stories were often "entangled" (Lee 2006) in the physical and social acts of raising, harvesting, preserving/storing, and consuming food. To do these acts, then, meant transmitting stories, meanings and understandings that were intimately tied to these distinct practices. Global Food certainly has a story and it too is undeniably embodied. But

Figure 2.1   1917 advertisement for Sunkist Oranges

What a Stray Swarm of Bees Started Fifty-Odd Years Ago

ONE August day, in 1865, a swarm of bees passed over the jewelry shop of a young man in the little town of Medina, Ohio. Jokingly, the young man told one of his assistants he would give a dollar for the bees. The man rushed out and in a few minutes returned with the bees all snugly housed in a light grocery box. The young man paid the dollar, took the bees and, being a lover of nature, at once became intensely interested in his new friends. It was the beginning. From that day this man and his family have devoted the best part of their lives to the study of bees, their lives, varieties, habits, culture and housing and to the betterment of their wonderful product, honey, through improvement of bee colonies and their surroundings. For his well-proved theory always has been that if you improve the worker and the worker's living conditions you're bound to improve the product.

The name of that man is A. I. Root, who stands today as the head and father of the bee industry in America. Recognized and quoted as an authority on bees and honey in that great work the Encyclopedia Britannica. In their long article on bees, reference after reference is made to A. I. Root, the A. I. Root Company and to Mr. Root's book on bees entitled "A. B. C. of Bee Culture." These references and illustrations are to be found in Volume 3, Eleventh Edition of the Encyclopedia; pages 630-632-633-634-636 and on page 638, where his name comes under the heading, "authorities."

Today, the house founded by A. I. Root and his stray swarm of bees is acknowledged to be the biggest producer of and dealer in bees, bee products and bee-keeper's supplies in the world. It is this firm, Mr. Root's pride and pleasure — a house maintaining firmly the ideals of its founder and operated entirely by his family—that offers you and guarantees to you the quality and purity of Airline Honey.

Though honey has always been used by many housewives in preference to sugar, the present situation has done much to teach vast numbers of others the superiority of honey.

The war has raised the price of sugar until it is an item well worth considering.

Honey makes you partially independent of sugar. Not only is it cheaper than sugar, but it is better for many things you're in the habit of using sugar for. It's more wholesome, it is more delicious and it gives better results.

Honey makes the finest, clearest preserves you ever tasted, and preserves the fruit better.

There never was such delicious cake, or cake that stays moist so long as that made with honey — nearly all kinds of cakes and cookies.

And candy! Just try some candy made with honey and you'll always use honey in making candy thereafter.

Spread honey on your batter cakes, waffles, bread and biscuits — there's no flavor in the world like it.

Use Airline Honey for many things in the way of cooking and sweetening and you'll rejoice in your discovery of honey's superior goodness — you'll benefit by its superior wholesomeness and profit by its economy.

You Need This Cook Book

It contains over 100 recipes for things in which Airline Honey is used instead of sugar. Also general directions giving the amount of honey to be used in replacing sugar as a sweetening, for one uses less honey than sugar. Send us your grocer's name and the book will go to you, free.

Send for Our Trial Jar

of Airline Honey. A generous quantity of the extracted honey in an individual jar. Send 10 cents for this treat—it will make you a honey enthusiast.

We Have Some Candy for You

—Delicious honey candy—several kinds —neatly and substantially packed. Ten cents in stamps to pay packing and postage will get you this candy. Send for it.

You can buy Airline Comb Honey in air-tight packages or extracted in glass jars (several sizes) with patent easily removable tops, at good grocers. Served in individual packages on most all railroad dining cars, at leading hotels and restaurants. This in itself is a striking endorsement of Airline quality—these people seek only the best and purest.

THE A. I. ROOT CO. Medina, Ohio
"The Home of the Honey Bees"

Figure 2.2   1917 advertisement for Airline Honey

as our physical relationship to food has changed so too have these stories. Many people, after all, still go to the grocery store looking for a good story. Perhaps it is a tale of how many miles a food has traveled before reaching one's dinner table (e.g., food miles labels). Or the story of the organic label, telling consumers about how a particular commodity was raised and processed. Or, following the a-picture-is-worth-a-thousand-words logic, some consumers may merely be looking for a friendly face—an Uncle Ben or a Betty Crocker—that they've known since childhood.

A lot has been written about how our food system has been hijacked by excessive rationalization (see e.g., Fantasia 1995; Ritzer 2003; Wright and Lund 2003). Following this thesis, products and processes have become standardized—and externalities (to ecological and social systems) largely ignored—to achieve maximum efficiency. This "squeezing" of the environment, humans, and non-human animals is allowed to occur due to a relational consequence of industrial, large-scale food. The *global* nature of today's food system also injects a level of "epistemic distance" (Carolan 2006c) into the experiential horizon of consumers that was unknown to previous generations. Some analysts contend that consumers derive a level of comfort from having their grocery store aisles and restaurant menus populated with artifacts with remarkable similarity (Ritzer 2008: 14–15). Perhaps; though it's hard to discern the arrow of causality in this case—in other words, did this psychological comfort lead to standardization or did consumers become tuned to standardization *after* the fact. A less controversial point of this rationalization thesis is how it creates a system in need of differentiation. And so we're back discussing the importance of the story. If the dominant food system is controlled by a handful of global corporate giants (Murphy 2008) and if, to achieve economies of scale, it works to expose our palates to relatively standardized flavors and sensations (Caldwell 2004) then the story becomes an indispensable part of what industrial food *is*. Stories seek to fill the holes—epistemic, emotional, sensorial, and otherwise—created by rationalization.

The French and Canned Food

France is well known as a site of resistance to industrial foods and the taste matrix it rests upon. The Slow Food Movement has a major presence there. France also has the Appellation d'Origine Contrôlée (AOC) ("controlled designation of origin"). Originally applicable to wines, today the AOC can be award to a variety of agricultural products (cheese, olive oil, etc.). Based upon a rigorous application process, AOC products must be produced in a manner consistent with artisan (rather than industrial) practices in designated geographical areas. This view that the French, as a whole, are somehow more tuned in to the tastes of non-industrial food items than, say, those living in the US is eloquently argued by Amy Trubek (2008) in her book *A Taste of Place*. Writing on the importance of local food systems in the US, Trubek looks to the French for inspiration. Her attention is directed particularly at the success they have had in developing what she calls a taste of place, or what the French call *le goût du terroir*:

> It is difficult to translate *terroir* from the French in a way that encapsulates all its meanings […]. It is part of people's everyday assumptions about food; it is as fundamental as our assumption that the first meal of the day should include coffee and orange juice but not miso soup. The French are unusual in the attention they place on the role of the natural world in the *taste* of food and drink. When the French take a bite of cheese or a sip of wine, they taste the earth: rock, grass, hillside, valley, plateau. They ingest nature, and this taste signifies pleasure, a desirable good. Gustatory pleasure and the evocative possibilities of taste are intertwined in the French fidelity to the taste of place (Trubek 2008: 9).

To be sure, there are aspects of this quote that make me uncomfortable. I get what Trubek is trying to convey, though I believe she is greatly overstating the point. It's her universalization of "*the*" French palate—as if the French all know food in the same way—that I have the most problem with. Criticisms aside, the argument is not terribly unusual. Plenty has been written on how the French seem to be more attuned to non industrial, artisan foods than residents of other Western nations (see e.g., Petrini and McCuaig 2004; Kummer et al. 2008).

I mention this not to engage in a debate about whether the French are somehow more tuned to non-industrial foods than, say, the average US citizen; after all, my research took place in the US, not France. I do think it is important, however, that we not essentialize the French palate and hence French embodiments towards food. Regardless of how people in France relate to food today that relationship—and the "tastes" rooted therein—is not immutable and fixed. The historical record is clear: the French palate, like palates embedded in all cultures (see e.g., Diner 2001; McWilliams 2005), has been in continual flux over the centuries (see e.g., Ferguson 2004; Freidberg 2004). I want to turn attention briefly to one example of this fluidity: the tuning of French consumers to canned foods. What's interesting about this case is that it shows how even the French body, long exalted for its ability to discriminate, has itself become tuned to industrial food. This tuning is the artifact of two movements: getting bodies to not only accept the practical requirements associated with industrial foods but also, ultimately, getting those bodies to actually "choose" this form of food over others.

As historian Martin Bruegel (2002: 113) notes, previous to the First World War there was a "lack of taste" for canned foods in France. While canning industries in other countries (most notably the US, Germany, and England) were experiencing steady growth, as argued by the editor of the *Journal d'Agriculture Pratique* in 1905, the French needed "to overcome the irrational as well as instinctive repugnance [for canned food] among a large part of the population. It would be an arduous task" (as quoted in Bruegel 2002: 113). The French, in a word, had to be *tuned* to the tastes and practical requirements of canned foods. Recognizing that I am throwing the ideology of consumer sovereignty out the window, bodies had to be conditioned to "choose" these foods.

Some resistance to these foods early on stemmed from health concerns. Defective cans could result in lead poisoning and botulism. One army survey records approximately 100 cases a year between 1886 and 1904 of canned food related illnesses (Bruegel 2002: 118). But improvements to the canning processes did nothing to make people like or even trust canned foods because trust first requires experience to build upon. Detailing how the French were slowly tuned to canned foods (though never using the term), Bruegel points to social institutions and practices. In his own words:

> The abatement of distrust and resistance resulted neither from a hypothetical word-of-mouth persuasion between individuals, nor from the enticement of chimerical status considerations. Because consumption is an eminently social phenomenon, social institutions and processes must suffice to explain the effects of new commodities in people's lives. The acquisition of a new taste requires the intervention of social groups physically to acquaint the newcomer with the merchandise and the technique required to manipulate it, to instigate and organized the first trial, to keep on pressure after the initial rejection, with explanations on the appropriate use of the good, and eventually to sustain routine consumption (Bruegel 2002: 118).

Bruegel proceeds to detail the role that the military and public schools played in this tuning process, quoting an Officer who declared in 1907 that "[t]he army is the continuation of the school". The army regularly exposed young French men to canned meals, as mandated by military rules. Indoctrination was also enhanced by the fact that soldiers, unlike, say, a citizen in a restaurant, had to eat what they were served. This conditioned solders to the taste of canned food and accustomed them to the practices of preparation—just opening the can and getting at the food, for example, were major practical hurdles to adoption that once overcome made the foods more acceptable. Regularly exposing soldiers to canned food also helped industrialists adjust their product to the "needs" and "tastes" of the consumer, recognizing how soldiers during this period came from all strata of French society and thus represented an ideal test (and tuning) population. Hence, when soldiers complained how upon opening a can of meat they were greeted by the unappetizing sight of a "lonely piece of meat swimming in a tasteless sauce" (Bruegel 2002: 123) manufacturers added carrots to the canned beef's contents.

The tuning process was enhanced with the coming of the Great War. The public was told to "add a few cans of food to every parcel" sent to soldiers on the front line (Bruegel 2002: 124). Cans of food purchased to feed the front line were often sold at discounted prices. Getting *citizens* to go to the store to buy canned food "proved the last threshold on the way to the quotidian purchase of sterilized comestibles" (Bruegel 2002: 124). The war also served to create new memories for this otherwise non-memorable food item. Canned food began to call forth memories among soldiers of time around the camp fire when food was eaten and shared with one's "brothers"; a moment of relative stillness in a period of their lives otherwise filled with pain and hardship.

As in other countries, industrial food processors also knew an additional method to convince consumers to purchase their goods was product tastings, which extended their reach to non-soldiering bodies. Food halls at fairs, in-store tastings, in-home demonstrations, and traveling railcars giving away free product gave consumers the opportunity to experience the food and (perhaps more importantly) begin to learn the practices associated with its preparation and consumption. Meanwhile, producers continually sought ways to improve the palatability of the food; not necessarily to make the foods taste better than homemade (or fresh) but to make them taste "good enough" so consumers would purchase them again for their convenience (it's immensely easier to buy a can of corned beef than to raise, feed, and slaughter a cow), durability (as in the case of canned versus fresh vegetables), or price (especially when looking to purchase a commodity out-of-season) (Fitzgerald and Petrick 2008: 399). Industrial food processors were interested, at least initially, in merely getting their foot in the door—or more accurately the mouth—of consumers; to get them accustomed to the novel practices and visceral experiences associated with industrial food. Recognizing that it takes time to tune a body to novel food-ways, the French canning industry realized they could not just flip a switch and make people want canned foods.

By the 1920s the canning industry's efforts were beginning to pay off. The French were slowly showing signs that they actually *desired* canned foods. In the words of some correspondents immediately following the war:

> Our soldiers brought back recipes from other regions and the habit of eating canned food (Bruegel 2002: 125);

> During the war, many a man left the country [*le pays*], saw new parts of the world, ate differently. Upon returning, they noticed that '*le pays*' lacks some goods to make life more comfortable—wine, white bread, canned foods—and so they do not hesitate to order a barrel of wine, to purchase a crate of pasta or to make ample provision of canned sardines (Bruegel 2002: 125);

> Since the Great War groceries have invaded the countryside: sugar, coffee, chocolate, pasta. Canned fished, canned vegetables used to be rare, today they appear even in the humblest families (Bruegel 2002: 125–6).

Though yearly per capita consumption of canned food in France lagged behind comparable industrialized nations at the dawn of the twentieth century, by the 1920s the average French mouth (and body) appeared to be as tuned to canned foods as their European neighbors. This change in consumption can neither be explained by something inherent to canned food. Nor can it be attributed to purely social-psychological accounts of "diffusion". It took work to make these bodies accept—indeed even *choose*—canned foods.

The Political Economy of "Tuning": The Soybean

As should by now be clear, our being tuned to Global Food did not just happen. We saw this in the case of France and canned foods during the early decades of the twentieth century; it took a tremendous amount of effort, coordination, and (non-market) intervention before bodies "chose" these industrial foodstuffs. I would like to spend a little more time locating this process of being tuned to Global Food within broader political economic forces. The purpose of this is two-fold. First, it demonstrates that a focus on lived experience need not compete with nor occur in the absence of a political economy approach. Secondly, showing the role of non-market forces in creating consumer desires and preferences—in a word, their *tastes*—that favor industrial foods gives force to those calls for non-market (e.g., state) support of alternative food systems.

The case I will now look at differs considerably from that of canned food. I shall briefly describe some of the political economic forces that helped tune US citizens to the soybean. While the US is the world's number one producer of this agricultural commodity its citizens have been slow to accept the soybean as a food (even though it has been a food staple in other countries for centuries). The slow acceptance of soybeans as a food item in the US is the product of a lot of work, money, and political power.

The story begins (at least as I tell it) in the 1920s with someone by the name of Henry Ford. In addition to being interested in its industrial applications, Ford believed the soybean would one day become a staple of the Western diet. In 1926 he hired a long-time friend, Dr Edsel Ruddiman, to experiment with the soybean. His instructions were to develop a biscuit. Ford ate these soybean biscuits regularly. He even claimed to like them (Wik 1962: 253). Visitors to the Ford estate, however, found the flavor of the biscuits less than pleasant. One of Ford's personal assistants later called the vitamin fortified soybean biscuits the "most vile thing ever put into human mouths" (as quoted in Wik 1962: 253). Another one of Dr Ruddiman's inventions was soybean milk, which he manufactured by rubbing beans together in the presence of water. This produced a high protein liquid. Other "constituents" were then added to give the liquid a comparable texture and viscosity to cow's milk. To promote these new foods, Ford would have guests over for dinner and serve them a completely soybean-based meal, from soybean soup, to soybean bread, soybean croquettes, soybean pie, soybean coffee, and soybean ice cream (Wik 1962: 253).

The dominant taste matrices of the time, however, were working against Ford and the soybean. Americans complained that soybean oil and protein were "bitter". The industry therefore worked to adapt to US tastes, rather than wait for palates to change. Advances in soybean processing after the Second World War improved the taste of soybean oil considerably. As the President of the German Society of

Fats and Oils Research explained at a Conference on Soybean Protein in 1973: "The problem is to fit soybeans to the taste of the white man [sic] who has still the taste of meat of his tongue" (as quoted in Berlan et al. 1977: 412). Even though soybean oil is today the most important edible fat in the US it is still rarely labeled as "soybean oil"; "persuasive evidence", some suggest, "of the wide-spread, if unspoken, American perception of soybeans as a nonfood item, more suitable for industry and animal feed than for humans" (Mintz et al. 2008: 5).

The American Soybean Association aggressively lobbied to bring the soybean to the classroom (or more accurately the lunchroom). They got their wish in 1971, as soybeans were inserted on the list of commodities supported by the National School Lunch Program. The goal of introducing soy protein to the still forming palates of American school children was to socialize taste buds to the soybean. It has long been known that taste preference is shaped considerably during childhood (see e.g., Birch 1980). Thus, in the words of former Secretary of Agriculture, Earl Butz, "a major break is realized in 1971 when the USDA authorized the use of textured vegetable protein enriched with vitamins and minerals in school meal programs" (as quoted in Berlan 1977: 413). While believed unrealistic to expect already-formed adult palates to change substantially, hence the push to "improve" the flavor of soy oils and proteins after the Second World War, the highly malleable taste buds of children were viewed as "open" to the soybean. Tune those taste buds to soybean-based foods at a young age, so the logic goes, and you will later have an adult population more likely to be soybean consumers (see e.g., Kauffman 1999: 417). In the first year that soy proteins were allowed into the program, US schools served 23 million pounds of vegetable protein, the vast majority of which came from the soybean (Johnson et al. 1992: 439–41; see also McCloud 1974). By the mid-1990s, soy protein constituted approximately 30 percent by weight of meat products served through the National School Lunch Program, which translates into over 50 million pounds of soy protein per year (Ensminger 1993: 20–24).

The decade of the 1970s also marked a moment when soy protein was finding its way in significant quantities into the American home, as a "meat extender". High beef prices in the early 1970s, due to high grain prices, caused grocery stores and schools around the country to "cut" their meat with soy protein. A school administrator in Illinois, in an interview for an issue of *Popular Mechanics* in 1974, reported that this practice saved the school "about 10 cents a pound" (Peterson 1974: 87). In the same issue of *Popular Mechanics* a food scientist is quoted as saying: "I expect that someday you may be eating soybean products because you like soybeans. For the present, soybeans will be used mainly for extending meat and producing look-alikes for meat favorites" (Peterson 1974: 188). How right this individual was. By the mid-1970s, as beef prices dropped, so did the practice of extending meat with soy protein; in fact processors felt the need to assure consumers that their product contained no "fillers" or "cereal additives" (DuBois 2008: 220).

The evidence suggests, however, that bodies were slowly becoming tuned to soy-based foods by the late 1970s. Health movements were gaining traction in the 1970s, many of which touted the benefits of the soy-based diet. In 1978 the Soyfoods Association of North America was founded, along with its magazine *Soyfoods*. By 1984, almost 200 companies in the US were producing tofu. An additional 53 made tempeh, 29 made soymilk, 21 made soy sprouts, 12 made miso, and 10 made soynuts (Shurtleff and Aoyagi 2001: 26). A 1977 US Gallup poll revealed a significant level of positive public awareness toward soy-based foods. Among the 1,543 adults interviewed, 33 percent believed that soybeans would be the most important source of protein in the future ("fish" and "meat" scored a 24 percent and 21 percent respectively). Fifty-five percent also agreed that "soy products have a nutritional value equal or superior to meat". The most positive responses toward soy came from younger respondents living in urban centers with college educations (Shurtleff and Aoyagi 2001: 26).

And today, the turn towards soy-based food continues. More than just feed for livestock, the soybean is becoming, for more and more North Americans, a food. Just a couple of examples of how our tastes and gastronomical temperaments toward the soybean are changing: soy milk is on the verge of becoming a $1 billion a year industry in the US (Yeong 2009); frozen edamame imports have increased from approximately 300 tons a year in the mid-1980s to about 25,000 tons a year by 2005 (Mentreddy et al. 2002); and, in 2008, 70 percent of all edible oil consumed in the US was derived from the soybean (Sutter 2009). The symbolic and cultural meanings attached to the soybean are clearly changing, creating new (and expanding old) channels into the food industry to levels never before seen.

Detailing the specific role that the soybean crushing industry has played in this process of tuning is complicated and would require more space than I have to devote to this discussion. The soybean agro-food/feed chain is remarkably concentrated. An analysis performed in 2002, for example, calculates that ADM, Cargill, Bunge and Ag Processing Inc. (AGP) controlled 80 percent of the soybean crushing industry (Hendrickson and Heffernan 2002). This level of economic power is not inconsequential in the formation of taste regimes. That said—and complicating the matter—edible or food-grade soybeans are an entirely different artifact from those crushed in ADM, Cargill, and AGP facilities. Bred to be larger-seeded, milder-tasting, more tender, and more digestible (they contain a lower percentage of gas-producing starches), soybeans raised for direct consumption (e.g., edamame) are about as similar to those found in fields throughout the Midwest as sweetcorn is to number two dent yellow field corn. Even from a production standpoint they are different, requiring special equipment due to their larger pods and seed. Approximately 90 percent of US food-grade soybeans in supermarket shelves are imported from Asia (Toland 2009); this, even though the US is the world's largest soybean exporter (29 million tons) (followed by Brazil with 17 million tons) (FAO 2006: 62). Yet placing the soybean crushing industry to the side it remains clear that political economic forces over the last half century have helped tune US bodies to soy-based food.

Recursivity and Recollecting

Memory is not located in the mind, or so a relational approach argues. If it were, memories would be as impossible to share as one's brain stem. Texts and images have been given special attention by scholars interested in the sociology of memory (see e.g., Connerton 1989; Halbwachs 1992). That is because words and images travel well. They are infinitely reproducible, tied down neither to any particular time nor place. As Rigney (2005: 20) explains: "Unlike material monuments, texts and images circulate and, in the process, they connect up people who, although they themselves never meet face-to-face, may nevertheless, thanks to stories and the media that carry them, come to share memories as members of 'imagined communities'". But this is a very "cold"—disembodied—understanding of memory. Sure, words and images are good to think with but our knowing of these textual "codes" (as they are often called by cultural scholars) does not occur in a disembodied vacuum. Building upon others' work (see e.g., Atkinson 2007; Terdiman 2003; Thrift 2004), I understand memory to be a process; a dynamic, shifting, and productive event rooted in a lived experience that cannot be reduced to words or an image.

The relationship between food and memory has recently attracted the attention of sociologists, geographers, and anthropologists. The contention is that the deeply visceral nature of food consumption makes food a particularly intense and compelling medium for recollection (Sutton 2001). The vivid memories of being forced to clean one's plate as a child (Batsell et al. 2002); the fading memories of traditional Greek culture as bodies adjust to the eating practices of a "modernity" (Seremetakis 1993); the performative nature of doing Korean culture among first-generation Koreans in Japan by consuming traditional dishes from Korea (Soo-jin Lee 2000): social scientists have been able to detail a variety of ways in which food is good to remember with.

Memories are also lived recursively, a point Foucault (2002: 239) was well aware of when he wrote on how knowledge and memories "are repeated, reproduced, and transformed; to which pre-established networks are adapted". Foucault was writing explicitly about declarations but his point could just as easily be applied to practices. For, ultimately, it is through recursivity—visiting the same places, repeating practices, experiencing visceral sensations over and over again—that memory is constructed and maintained (Rigney 2005: 20). If memory is something we *do* then what does this say about our daily trips to grocery stores and fast food places when it comes to knowing food? What sort of food memories do these practices create?

Connerton (1989: 23) argues that social scientists have neglected "habit memory", choosing instead to focus too exclusively on more personal and cognitive forms of memory. Connerton is not breaking any new ground by noting how memory has a recursive quality to it. Henri Bergson made a similar distinction a century ago between memory that consists of habit and that which consists of recollection. Bergson uses the example of learning a music lesson by heart. For Bergson (2007: 95), while we say we "remember" the lesson it is more accurately "habit interpreted by memory rather than memory itself". Contrast this to remembering the first time the musical score was read, which represents for Bergson (2007: 95) "memory *par excellence*". The novelty of Connerton's argument, rather, lies in his audience: social scientists. I agree with Connerton that more attention needs to be placed on unpacking how memories are *performed*. A ritual, for instance, which lays at the heart of most memory events, "is not just a symbolic representation of something else, not simply a text that needs decoding, but a matter of bodily repetition that is both performative and mnemonic" (Spillman and Conway 2007: 84). But I still question whether the (analytic) distinction between cognitive and habit memory actually exists. Isn't memory, all memory, to some degree performance dependant?

To go back to Bergon's example: wouldn't that memory of when the musical score was first read be imbued with performative elements? Elements like: was the musical lesson successfully mastered; in what manner was this mastery attained (that is, was the pupil forced to practice daily for hours on end); is any physical pain or discomfort associated with the memory (for example, do they have arthritis or were they practicing on an unpadded seat)? And so it is with food, which brings us back to knowing Global Food through our regular trips to grocery stores and fast food places.

Grocery store chains, food manufacturers, and restaurant chains: all work hard to make the consumer's experience pleasant (if not downright enjoyable). Grocery stores make their fresh fruit and vegetable displays (which are often located at the very front of the store) viscerally arresting so as to attract bodies in and keep them there as long as possible (Freidberg 2009: 123). Marketing research shows that if a grocery store pumps in the smell of baked goods, sales in that department increase (in some cases a three-fold increase has been recorded) (Lamb, Hair, McDaniel 2008: 383). Natural lighting has also been shown to increase sales in grocery stores (Lamb, Hair, McDaniel 2008: 383). And for children, so their mnemonic experience is also a positive one, there are free cookies, shopping charts molded in the shape of trains and race cars, and, in places like McDonald's, clowns and play areas.

The production of extra-local memories has been important to the mobilization of what Anderson (1991) terms an "imagined community". Anderson uses the term to describe how the categories of "us" and "them" are produced (and continually reproduced) through nationhood, emphasizing specifically the recursivity that goes into doing nationalism. Often these memories serve powerful interests. Connerton (1989: 51) argues that "it is now abundantly clear that in the modern period national elites have invented rituals that claim continuity with an appropriate historic past, organizing ceremonies/parades and mass gatherings, and constructing new ritual spaces". Connerton was not thinking about food—or the food system—when writing on how societies remember. Nevertheless, parallels exist between the memory making abilities of elites in the making of a nation's history and the memory making abilities of Global Food. The food populating our grocery stores try hard to claim continuity with the past. Claims like "all natural" and "freshly squeezed"; the friendly faces talked about earlier that grace the boxes and cans of today's foods; the stories told about "happy cows" grazing in pastures: all are attempts to evoke memories and in so doing make industrial food, literally, *event*-full.

Remember back to the case of canned foods in France. One early barrier to bodies becoming tuned to these canned artifacts was that this food was literally too industrial. It was cold, mass produced, and unfamiliar. It lacked, in a word, memories. Memories make foods unambiguous—they help form feelings of "like" or "dislike" toward a consumer good. At least according to one historical account (Bruegel 2002), the First World War created important (positive) memories toward canned foods. These visceral memories, as mentioned, were of relaxing around a campfire with one's unit. Canned foods after the Great War, at least for some French veterans, evoked sensations of brotherhood and camaraderie during a moment when the world was otherwise uncertain. Part of this tuning process to industrial food, I believe, is the development of memories of these foods. Memories as much as taste keep us coming back for more. I now prefer fresh mushrooms over canned. But I must admit: whenever I eat canned mushrooms I am awash with fond memories and sensations from when I was a kid. I could probably rattle off more than a dozen industrial food items—from Mr. Freeze Freeze Pops, to Push-ups, Chef Boyardee canned foods, and Little Debbie snack cakes—that take me back in a very visceral, positive way my childhood. Those sensations may get me to eat some of those foods again in future, even though I know I shouldn't.

Global Food and Distance

Though rarely mentioned in the agro-food studies literature, Martin Heidegger offers some useful concepts to help us make further sense of our relationship to food, particularly as that relation has become "stretched" with the lengthening of agro-food chains. Heidegger (2000: 45–50) makes the distinction between "readiness-to-hand" (*zuhandenheit* [artifacts-in-use]) and "present-at-hand" (*vorhandenheit* [artifacts asking attention for themselves]). He uses these terms specifically to talk about two modes of human-technology relations. Heidegger argues that technological artifacts tend to draw attention to themselves only when they break down. When nailing something to the wall, for example, our attention is drawn to the nail. Only when the hammer slips or breaks does it come to the foreground of cognition. Similarly, while we may directly stare into the television our attention is actually drawn to the images and sounds of the movie being watched. Only when the television does not turn on or after the DVD skips do these artifacts call attention to themselves. When functioning properly, the television can be said to have a readiness-to-hand quality as it withdraws from our perceptual field and becomes (to a degree) transparent. When malfunctioning, the television enters into a different relation with us, becoming present-at-hand as it ceases to be as transparent.

It is important not to speak too exclusively when utilizing these terms. We must realize that technological artifacts exist in a type of Gestalt form. Nothing is ever in a pure state of either readiness-to-hand or present-at-hand because the technological artifact's state of being changes with the conditions of use. Going back to the television example, the artifact brought into a present-at-hand state varies according to the malfunction. For example, a "burned-in" image will bring to the foreground the television's plasma technology whereas the sound of crackling/ distortion will cause the television's factory speakers to lose their transparency. I mention this point to remind ourselves that nothing can ever be entirely known. Even though, for example, I am very aware of my laptop—not only do I stare at it for hours every day but I have also become quite familiar with its feel, sounds, and smells—there remain aspects of it that have a readiness-to-hand quality for me, such as its motherboard, its underside, and all that went into its manufacturing.

As a sociologist, I am bothered by Heidegger's apparent lack of interest in *why* artifacts take on a readiness-to-hand quality. Heidegger, you see, was speaking to the ghosts of other philosophers. For Heidegger, Western thought had long been concerned with presence at the expense of the absence that saturates our world. Heidegger's originality lies in this break from ghosts past, by emphasizing just how much of the world floats by unnoticed and unrecognized. But while a world-class philosopher, Heidegger wasn't much of a sociologist. Heidegger is correct to say that these readiness-to-hand qualities emerge through habitual actions; through our relational state to the world (what Heidegger awkwardly called our being-in-the-world). But these relational states do not just happen. To put it bluntly: there are sociological reasons why we have the habits that we do and thus reasons why the absences and presences of the world are what they are.

Let's apply this thinking to food. There are many aspects of the food system that are not easily perceived at the point of consumption. Environmental degradation, social injustices, and reductions in biodiversity are just a few of the more publicized absences associated with Global Food. But there are other absences, too, like forgotten cultural memories and practices, animals, and certain tastes. Such absences, I argue, make it easier for this system of food provision to continue. In the chapters that follow I discuss these absences and their role in maintaining—and subverting when turned into presences—the conventional food system. To be sure, all food has readiness-to-hand qualities. I am not even trying to suggest that alternative food systems come with fewer readiness-to-hand qualities than Global Food. My point, rather, is that, if we look closely, the readiness-to-hand qualities of Global Food obscure those very things that could, if made present-to-hand, threaten its existence.

Chapter 3
Making Sense with CSAs

I recently came across an interesting piece of mapping wizardry. Developed by Stephen Worley, the map was constructed by taking the locations of over 13,000 McDonald's in the US and plugging them into some special software. The end product is a map of the lower-48 that plots each McDonald's, represented by a bright pinpoint against a black background. East of the Missouri River is almost solid light, save for small spots of darkness signifying the Adirondacks, inland Maine, the Everglades, and the hinterlands of West Virginia. West of the Missouri River—especially around central Nevada, southeastern Oregon, Idaho's Salmon River Mountains, and the high plains of South Dakota, Wyoming, and Montana—one will have to travel a little further between Golden Arches. And the spot furthest from the Big Mac with its two all beef patties, special sauce, lettuce, pickles, and cheese? You'll have to go to South Dakota to find it, between the small towns of Meadow and Glad Valley. The "McFarthest Spot" in the lower-48 is 107 miles (as the crow flies) from the nearest McDonald's, 145 miles if one stuck to the roads (Worley 2009).

Staring at this map brought to mind a far less sophisticated project that I began over 10 years ago. While a graduate student at Iowa State University I attempted to "map"—using the term loosely here—the changing commodity profiles of farms for a handful of states in the Midwest. With each new census of agriculture—the most recent, for the year 2007, was released on February 4, 2009—I have continued to update those tables. Given that the empirical material for this chapter was gathered in Iowa, I will reproduce the state's table here (see Table 3.1). Note the state's shrinking commodity basket. In 1920, 34 commodities were produced on at least 1 percent of Iowa farms, while 10 commodities were produced on at least 50 percent. In 2007, those numbers shrank to nine and one respectively.

I am the first to admit that empirical maps like these, while fun (or for some scary) to look at, don't really tell us much beyond what they claim to. And certainly, when it comes to the subject of embodied knowledge, such figures are like flying at 30,000 feet in terms of conveying an understanding of the lived experience of those whose feet are planted firmly on the ground. Nevertheless, views from this height do have their place. Before satellite imagery technology petroleum geologists used to map vast areas with a helicopter searching for geological formations that might yield oil—to tell them, in other words, where to dig. Similarly, looking at the aforementioned McMap and commodity profile table compels me to drill down and learn more about the trends they rather superficially illustrate, which, in this case, involve further understanding of how we know food. Commodities in bold indicate being produced on at least 50 percent of all farms.

Table 3.1   Number of commodities produced for sale in at least 1 percent of all Iowa farms for various years from 1920 to 2007

1920	(%)	1935	(%)	1945	(%)	1954	(%)	1964	(%)
Horses	**(95)**	**Cattle**	**(94)**	**Cattle**	**(92)**	**Corn**	**(91)**	**Corn**	**(87)**
Cattle	**(95)**	**Horse**	**(93)**	**Chicken**	**(91)**	**Cattle**	**(89)**	**Cattle**	**(81)**
Chicken	**(95)**	**Chicken**	**(93)**	**Corn**	**(91)**	**Oats**	**(83)**	**Hogs**	**(69)**
Corn	**(94)**	**Corn**	**(90)**	**Horses**	**(84)**	**Chicken**	**(82)**	**Hay**	**(62)**
Hogs	**(89)**	**Hogs**	**(83)**	**Hogs**	**(81)**	**Hogs**	**(79)**	**Soybeans**	**(57)**
Apples	**(84)**	**Hay**	**(82)**	**Hay**	**(80)**	**Hay**	**(72)**	**Oats**	**(57)**
Hay	**(82)**	**Potatoes**	**(64)**	**Oats**	**(74)**	Horses	(42)	Chicken	(48)
Oats	**(81)**	**Apples**	**(56)**	Apples	(41)	Soybeans	(37)	Horses	(26)
Potatoes	**(62)**	**Oats**	**(52)**	Soybeans	(40)	Potatoes	(18)	Sheep	(17)
Cherries	**(57)**	Cherries	(34)	Grapes	(23)	Sheep	(16)	Potatoes	(06)
Wheat	(36)	Grapes	(28)	Potatoes	(23)	Ducks	(05)	Wheat	(03)
Plums	(29)	Plums	(28)	Cherries	(20)	Apples	(05)	Sorghum	(02)
Grapes	(28)	Sheep	(21)	Peaches	(16)	Cherries	(04)	Rdclover	(02)
Ducks	(18)	Peaches	(16)	Sheep	(16)	Peaches	(04)	Apples	(02)
Geese	(18)	Pears	(16)	Plums	(15)	Goats	(04)	Ducks	(02)
Stwberry	(17)	Mules	(13)	Pears	(13)	Grapes	(03)	Goats	(02)
Pears	(17)	Ducks	(12)	Rdclover	(10)	Pears	(03)	Geese	(01)
Mules	(14)	Wheat	(12)	Mules	(06)	Plums	(03)		
Sheep	(14)	Geese	(11)	Stwberry	(06)	Wheat	(03)		
Timothy	(10)	Sorghum	(09)	Ducks	(06)	Rdclover	(03)		
Peaches	(09)	Barley	(09)	Wheat	(04)	Geese	(03)		
Bees	(09)	Rdclover	(09)	Timothy	(04)	Popcorn	(02)		
Barley	(09)	Stwberry	(08)	Geese	(03)	Timothy	(02)		
Raspbry	(07)	Soybeans	(08)	Rye	(02)	Swtpatoe	(02)		
Turkeys	(07)	Raspbry	(06)	Popcorn	(02)	Swtcorn	(01)		
Wtmelon	(06)	Bees	(05)	Swtcorn	(02)	Turkeys	(01)		
Sorghum	(06)	Timothy	(05)	Raspbry	(02)				
Goosebry	(03)	Turkeys	(04)	Bees	(02)				
Swtcorn	(02)	Rye	(02)	Sorghum	(01)				
Apricots	(02)	Popcorn	(02)						
Tomatoes	(02)	Swtcorn	(02)						
Cabbage	(01)	Swtclovr	(01)						
Popcorn	(01)	Goats	(01)						
Currents	(01)								
n = 34		n = 33		n = 29		n = 26		n = 17	

Table 3.1   (Continued)

1978	(%)	1987	(%)	1997	(%)	2002	(%)	2007	(%)
Corn	**(90)**	**Corn**	**(79)**	**Corn**	**(68)**	**Corn**	**(58)**	**Corn**	**(54)**
Soybeans	**(68)**	**Soybeans**	**(65)**	**Soybeans**	**(62)**	**Soybeans**	**(54)**	Soybeans	(45)
Cattle	**(60)**	Cattle	(47)	Hay	(42)	Hay	(37)	Hay	(28)
Hay	**(56)**	Hay	(46)	Cattle	(42)	Cattle	(35)	Cattle	(32)
Hogs	**(50)**	Hogs	(35)	Hogs	(19)	Horses	(13)	Horses	(11)
Oats	(34)	Oats	(25)	Oats	(12)	Hogs	(11)	Hogs	(09)
Horses	(13)	Horses	(10)	Horses	(11)	Oats	(08)	Oats	(03)
Chicken	(09)	Sheep	(08)	Sheep	(04)	Sheep	(04)	Sheep	(04)
Sheep	(08)	Chicken	(05)	Chicken	(02)	Chicken	(02)	Goats	(02)
Wheat	(01)	Ducks	(01)	Goats	(01)				
Goats	(01)	Goats	(01)						
Ducks	(01)	Wheat	(01)						
n = 12		n = 12		n = 10		n = 9		n = 9	

Source: US Census of Agriculture, 1920–2007. Prepared by Michael S. Carolan, PhD; Department of Sociology; Colorado State University; Fort Collins, CO; 80526; mcarolan@colostate.edu.

Why Community Supported Agriculture?

Almost from the beginning of industrial food, attempts have been made to reduce the distance between the sites of food production and consumption. One of the earliest was the Nature-Study Movement, which sought to reintroduce students to experiences lost with urbanization and the industrialization of food production. The Nature-Study Movement reached its peak in the late 1800s and early 1900s. In 1897, for example, over 26,000 students had "class" in school gardens in New York State (Vileisis 2008: 106). The goal of the movement was to physically bring people into closer contact with nature by giving them firsthand knowledge of food production. As a then-seminal text on the subject explains, the movement was directed at attaining two ends: "to discover new truth for the purpose of increasing the sum of human knowledge; or to put the pupil in a sympathetic attitude toward nature for the purpose of increasing the joy of living" (Bailey 1903: 4). Later, the text mentions how children "should be taught something from the farmer's point of view" (p. 59). In another book it is announced that "[t]o allow a child to grow up without planting a seed or rearing a plant is a crime against civilized society" (Hodge 1902: 10).

The tradition of pragmatism gave this movement pedagogical and philosophical grounding.[1] John Dewey (1897: 77–80), for instance, was a vocal proponent of both the means and ends of the movement. Rather than viewing knowledge as this otherworldly thing that is accessed by the rational mind, pragmatism grounds knowledge in practice. Experience, therefore, and not abstract theories, marks the starting point for understanding. It makes sense why pragmatism provided fertile ground for the Nature-Study Movement. Contra the abstract frameworks and representations that are such a significant part of conventional curricula, pragmatism points out that there is more to know than what can be told to (or read by) students. While occasionally critiqued on the grounds of being anti-modern (Deloria 1999: 102), the movement's emphasis on artifacts like school gardens could be understood as an attempt to minimize society's growing detachment towards food and nature. It is interesting that this movement reached its height when it did, at the very moment when structural changes in agriculture were making the farming lived experience slowly a thing of the past.

But the distance between farm and fork continued to grow. Perhaps the short-lived Nature-Study movement was a gut response to a system of food provision among bodies not yet tuned to it. For example, attempts in the 1930s in the UK to keep food under glass to protect it from flies met considerable public resistance as shoppers still wished to physically handle their foods before making a purchase

---

1   It needs to be stated that, at least in some quarters, the nature-study pedagogy was grounded in the theory of recapitulation. Recapitulation was the belief that as humans developed from child to adult they repeated the evolutionary history of humans. Following this, children were believed to be cognitively similar to Native American Indians: primitive with an innate closeness to nature (Armitage 2007).

(Blay-Palmer 2008: 20). Eventually, however, these generations came around to accept the sensual requirements of this new food system. In 2001–2, for example, the global trade in processed food outstripped for the first time global trade of unprocessed agricultural products (Oosterveer 2007: 5).

In recent years calls have grown for less industrial food. Concerns over pesticide use, Mad Cow disease, biotechnology, the use of growth hormones in livestock and milk production, Escherichia coli (aka E coli), and the environmental costs of industrial agriculture have expanded more "careful" forms of consumption in Western countries (Ilbery and Kneafsey 2000: 317). Food processors are responding by attempting to reduce the perceived distance between farm and fork, such as through packaging, labeling, and certification programs (Blay-Palmer 2008: 123; Renard 2003: 87–90). Other strategies involve creating shorter food chains, in the belief that greater first-hand knowledge will translate into fewer food fears and greater embedded trust in the producer-consumer relationship (Goodman 2003: 1–4; Hinrichs 2000: 296–9, 2003: 39–42; Watts et al. 2005: 29).

Some proponents of shorter food chains also claim that this food tastes better and that it is healthier than that provided by industrial means (e.g., Pollan 2008; Waters 2008). Yet, as others note (e.g., Freidberg 2009; Guthman 2002), such claims must be treated with caution. Taste, for example, has long been recognized among sociologists as having the potential to be a powerful class marker (see e.g., Bourdieu 1984; Elias 2000). And as Freideberg (2009: 2) thoroughly demonstrates, "freshness", as we know it today, is "dependent on a host of carefully coordinated technologies, from antifungal sprays to bottle caps to climate-controlled semi trucks". Moreover, as Allen and colleagues (2003) explain, the emphasis placed on taste and health by local and organic proponents has to some degree sidelined narratives that originally drove organic producers. For early organic growers and consumers, social and economic objectives motivated much of their behaviors, not the issues of taste and positivist definitions of food safety. Originally meant as an alternative system of food provision, organic food is in many cases becoming co-opted by the very interests that it initially sought to break free from: namely, those of Global Food.

A lot of ink has been spilt by social scientists examining local food networks (see e.g., Kneafset et al. 2008; Hinrichs 2000; Kirwan 2006; Lyson 2004; Selfa and Qazi 2005). These spaces—like CSA schemes and farmers' markets—have been shown repeatedly to represent more than a niche marketing strategy. It is this "more than" that sociologists and geographers have been drawn to. One need look no further than the writings of Granovetter (1985) and Polanyi (2001) to understand that all economic behavior is embedded in, and mediated by, social relations. Consequently, social "embeddedness should not be seen as the friendly antithesis of the market" because the market itself is a thoroughly social phenomena (Hinrichs 2000: 296). Agro-food studies scholars have been particularly interested in describing the type of social networks that exist in these spaces and their effects on exchange. Or, to put it another way, they want to know "to what degree embeddedness occurs and in what forms, and how a food system becomes embedded" (Milestad et al. 2010: 230).

Like my colleagues, I too am interested in this "more than" of local food systems. But I think spaces like CSA offer still more than what my colleagues give them credit for. I see no reason to dispute the argument that the spatially proximate social relationships found in these spaces (recognizing that spatial relations cannot be conflated with social relations [Hinrichs 2000]) can build trust between consumer and producer and thus help grease exchange by, among other things, creating consumer loyalty (Kirwan 2006). But perhaps there's more to embeddedness than just *social* embeddedness. That's what I was interested in finding out when I decided to study the CSA lived experience.

The Case: Introductions and Analysis

CSA consists of a group of individuals who pledge support to a farm or a group of farms. The goal of these organizations—at least one of them—is to create a local food system. Usually, members—or "share-holders"—pledge in advance to cover the anticipated costs for the year. CSA is therefore different from direct marketing because members commit to a full-season price in the spring and thus share in some of the risks that are inherent in farming. In return, members receive shares of what is produced for the year. In addition to receiving locally produced food, members are encouraged to visit the farms and walk through the gardens to experience firsthand where their food comes from. In short, CSA seeks to reduce the distance between consumers, growers, and food.

There are a number of works that detail the history and operational specifics of CSA (see for example DeLind and Ferguson 1999; Henderson and Van En 2007). To avoid discussing CSA in general historical terms, and merely repeat stories already told, I will focus on more novel matters by turning immediately to the empirical analysis. This analysis centers on two CSAs located in rural Iowa. In the two CSA schemes studied, consumers and growers met each other on a regular basis. Both also held "volunteer parties" a few times a year, which brought non-growers to the farm to help harvest. It is worth noting that this level of involvement between growers, consumers, and food production is somewhat unusual. As best as I can tell, the average CSA member never visits the farm from which their shares originate nor do they personally know the farmers responsible for growing their food. I mention this to caution the reader against generalizing these finding across all CSA schemes. It is to a specific CSA-type that this discussion pertains.

The research for this chapter was conducted in the summer of 2003. Two CSAs located in central Iowa (US) were examined. In total, 22 personal interviews were conducted, 18 were customers and four were farmers. Each interview lasted between 30 minutes and one hour. To further supplement the data, participant observation techniques were employed. This took the form of lending a hand gardening on these farms and delivering food to consumers at either predetermined distribution points or, for less mobile individuals, their homes (consumers could also come to the farm to pick up their share). All observations and interviews were transcribed, coded, and analyzed. The names of respondents have also been changed to protect their identity. I also tried to be aware of my own body when conducting this research. Though an unconventional form of data—most scientific analyses work so hard to deny the corporality of the analyst—I occasionally took the time to speak into a tape-recorder about my own experience of a space. I did this to inject another layer of sensuality into the text in the hope of giving the reader a better feel for the lived experiences described below.

Each CSA had a similar commodity profile, limited largely to vegetables and seasonal fruits like strawberries, watermelons, and cantaloupe. Both farms were approximately the same size, with roughly 1½ acres in "production" (this figure does not include the additional greenhouse space found on each farm). Perhaps the most noticeable difference between the two CSAs was that one had chickens (and thus offered eggs to its customers), while the other had no livestock. In all, the two farms were remarkably similar.

Most of the respondents described themselves as middle class. The remaining (n=4) felt the category lower-middle class better suited their socio-economic position. I spoke with 12 women, eight men. And all but two of the respondents identified themselves as Caucasian. The other two described themselves as African American and Asian American. Everyone I spoke to resided either in the countryside or a town (and no one lived in a town with more than 10,000 residents).

Making Food More Present

> It's your typical Iowa summer day, hot and very humid. I am on my knees pulling weeds. Corn stalks are all around me. The odor of rich organic soil is very distinct. With each weed I pull I think of all the nutrients I have freed up for the corn plant. Hand over hand, I move forward. Pulling. The removal of each weed, I tell myself, will make the kernels fuller, the corn sweeter. I know I'll think about this moment weeks from now when I am enjoying these ears with a little butter and salt (personal notes).

The industrial food system has helped to make food an object. As Marx would say, we have come to fetishize the thing-in-itself at the expense of relations when it comes to knowing what we eat. This is arguably best on display in surveys that reveal a surprising percentage of people who fail to recognize that milk comes from cows and bacon from pigs, and that pickles are cucumbers. To be sure, we still know food like a body. We always will, given how cognition is inherently embodied. But the *types* of embodied relations have changed, due in part to changes in the structure of agriculture over the last century.

I did eventually get to eat some of that corn I tended to. And there was something experientially different about it compared to, say, the ears I occasionally purchase from the local grocery store. It would not be an overstatement to say that the kernels were literally event-*full*. As I ate them I was continually reminded of my time in the garden. To be honest, with my lived experiences being what they are I rarely ever just see an ear of corn when I come into contact with this delicious food. I grew up around sweetcorn. My father grows approximately half an acre of this delicious vegetable every summer. He has been doing this for as long as I can remember. My past is rich with memories of pulling weeds between rows, in addition to picking corn, shucking it, boiling it, freezing it and, of course, eating it. So even when I buy corn-on-the-cob at the grocery store I still have a feel for more than the materiality (e.g., cellulose, starch, and sugar) before me. Yet my embodiments, being what they are, also cause me to understand other foods more objectively. Though I enjoy eating fish, for example, I admit to often giving little thought to the object wrapped in butcher's paper that I get from my store's refrigerated "fresh" fish section. Had my father been more interested in aquaculture, and less interested in sweetcorn, perhaps I would feel differently toward the salmon and halibut I enjoy eating.

Earlier I introduced some concepts provided by the twentieth-century philosopher Martin Heidegger (2000), that of "readiness-to-hand" and "present-at-hand". The former, to provide a quick reminder of the significance of these terms, refers to an artifact that exists as an absence, whereas the latter refers to those phenomena that we are aware of. With its industrialization, our relationship to the food system has come to resemble that of many of our technological artifacts. For most, the food system is highly invisible, save for the points of purchase and consumption (see Adam 1999). Our relationship with it has become one of a readiness-to-hand. Sometimes this changes, briefly. A food contamination scare can bring certain aspects of the system into momentary view (Kriflik and Yeatman 2005: 11–14). Yet, when properly functioning, the nature of the food system is such that we often forget about the goings on that take place before food reaches the store.

In additional to the food system, I would contend that food *itself*, at least in the US, has taken on certain readiness-to-hand qualities. This is one of the underlying themes of Michael Pollan's (2008) recent bestseller *In Defense of Food*. According to Pollan (2008: 7), "most of what we're consuming today is no longer, strictly speaking, food at all, and how we're consuming it—in the car, in front of the TV, and, increasingly, alone—is not really eating, at least not in the sense that civilization has long understood the term". As others have argued, "food" is not an objectively given category. Rather, is it "something that is performed, something enacted, and not something that necessarily demands rational, logical reasoning" (Roe 2006: 112). As Harris (1986: 13) reminds us, "[w]e can eat and digest everything from rancid mammary gland secretions to fungi to rocks (or cheese, mushrooms, and salt if you prefer euphemisms)". Pollan rather boldly tells the reader what he thinks is, and is not, food. No doubt this is a definition that hinges, at least in part, on Pollan's own lived experiences of food. I therefore understand why his definition of "food" differs significantly from, say, students that I have encountered that live off of Hot Pockets, protein bars, and Red Bull. These radically different definitions of food do not just happen. Nor can they be explained solely by the power of advertisements to inscribe upon individual subjectivities particular sentimentalities toward what we eat (Williamson 1983: 42), otherwise Pollan, unless he's been living under a rock all his life, would have the same grip on food as my Hot Pocket eating students.

"You know what I've got here, Mike?"

I was standing on the edge of a garden. The person in front of me was holding, in two hands clasped together, a bunch of green beans. I was asking him how he thought about food and how those understandings where shaped by his previous CSA experiences.

I answered, "Well Nick, they look like green beans".

"It was a trick question. Most people, when they think of vegetables, only think of that part which is edible. And why wouldn't they? When does the average person see a row of green beans or an asparagus bed? So, naturally, their mind will immediately go to what they know, which for most is the stuff you see at the store".

"So what is your understanding of what you're holding?", I asked.

"Well, green beans. I'm not hallucinating thinking there's something else in my hands. But I am also thinking about the plants of the green beans. You know, their roots are host to bacteria that fix nitrogen in the soil. I guess I'm trying to say how limited a view most people have of the food they eat. I really doubt that your average person, when picking up green beans at the store, thinks about the plant's nitrogen fixing qualities or how far those beans traveled so they could end up on their dinner plate".

The lived experiences offered at the CSAs studied brought issues of food production into a state of awareness, giving food a present-at-hand quality. That was Nick's point. Having the lived experience that he does gives him an understanding of green beans not available to your average person, who relies exclusively upon the conventional food system for their vegetables. Without actually utilizing the philosophical terminology, this was a common sentiment conveyed by respondents: that food has taken on qualities reminiscent of Heidegger's readiness-to-hand relation. By this I am not only saying that food production is becoming increasingly unknown by the average consumer. Many people go their entire lives without stepping on a farm or garden, knowing the tilth of rich, organic soil between their fingers. You do not have to interview anyone to figure this out. What I mean is that food *itself* is becoming transparent. I once overheard my aunt ask my then-16 year old cousin if she had eaten anything earlier in the day. When she asked this question it was 5:30 pm, almost time for supper. My cousin responded with a quick "No, I don't think so" as she passed through the kitchen on her away to her bedroom. When she came back a few minutes later to wash her hands she stopped suddenly to tell us how she just remembered that she and her friends did go through the drive-through to get French fries and ice cream after school. When eating fails to register in one's consciousness even when actively attempting to recall the act, as was the case for my cousin, I think an argument could be made that food is becoming increasingly an absence (versus a presence) in our lives.

A fair amount has been written about our instrumental treatment of food, which involves reducing food to, say, energy, or calories, or nutrients (Krasteva-Blagoeva 2008: 25–30; Kummer et al. 2008: 20–22; Petrini and McCuaig 2004: 16–26). Pollan refers to this phenomenon as "nutritionism". In Pollan's (2008: 28) words, "it is not the same as nutrition [...] as the 'ism' suggests, it is not a scientific object but an ideology"; an ideology rooted, I might add, in particular embodied practices. One individual eloquently explained how her experiences through CSA help reduce food's transparency:

> When you cook something, and take a long time to prepare it, you're not going to just shovel it down—right? Well it's the same when you know someone has taken the time to grow the food for you, especially if you've played a role yourself in this process. When I go to McDonald's—which I still do on occasion—it's just food. And I gobble it down without really thinking about. Not so with what you see around here [pointing around at the garden]. This isn't just fuel. This food is special precisely because it isn't just food.

This individual nicely contrasts conventional food with CSA food: the former is often reduced to a thing—to, in their words, "just food"—whereas the latter is something more. This "more" comes from the food having a present-to-hand quality to it, making it not just a means but an end itself. Another individual put it to me this way: "When all you know is grocery store food it's hard not to see right through it. You buy food-for-dinner or afterschool snacks for the kids. Learning to appreciate the food itself is not something we're use to doing anymore". While still another quipped: "I don't think as much about prepackaged or processed food when I eat it. I mean, beyond the immediate sensations of taste and smell I often don't give it much thought at all".

Hidden beneath all of this is a certain premium placed upon slowness. The concept of slowness has been talked about in recent years by social and political theorists as being an important strategy for resisting the quickening pace of everyday life (see, for instance, Connolly 2003; Mackenzie 2002; Stengers 2002; Wolin 1997). Slow food, slow living, and slow knowledge (to name but a few "slow" examples [see Honore 2004 for a more detailed list]): such "movements", at their heart, are about the search for more concrete, sensual experiences. In the case of the Slow Food Movement (see, e.g., Petrini and Watson 2001; Pentrini and

McCuaig 2004), the goal, or at least one of them, is to reclaim the "gastronomic aesthetic" (Miele and Murdoch 2002: 313). Respondents were interested in doing something similar when they talked about resisting the tendency to think about food as "just food".

It is also important to note that slowness, well, takes time. And time is not something everyone has. This is an important critique to keep in mind when talking about slow food: namely, that sometimes there are structural reasons why we're so fast with our food (Shteyngart 2006: 140–41). But I think some of this critique can be deflected as it applies to the cases studied in this chapter because locally grown food can play an important role in improving the food security of those living on the margins (Allen 2007: 45). So maybe a side-effect of building community food security is that we also won't have to be quite so fast with our food.

On Distinction

One of my interviewees was a self-proclaimed "strawberry connoisseur". "Give me a strawberry", he once told me, "and I can tell you the variety, soil type, and how long ago it was picked". Initially I thought he was just showboating for me (his personality was perhaps the most boisterous of anyone I interviewed for this research). But as we talked, I became convinced that he was onto something, even if his strawberry powers were being overstated (I never did get to test them). Talking more about his abilities he asked me, "Why does that seem strange? People do that with wine all the time and no one thinks it unusual. Why couldn't someone do that with strawberries or any other fruit or vegetable for that matter?" That is when he made the following insightful sociological point: "We still have that close relationship with wine and grapes so we still have that knowledge of them. We lost that with just about everything else [that we eat]".

I know someone that claims they can, with one bite, tell you exactly the region that a cantaloupe was raised in. For this person, there is no Cantaloupe, as a monolithic food category, but cantaloupe from the plains of Colorado, Pacific Northwest region, the Midwest, and so forth. I am fairly confident that our ancestors never possessed this knowledge either, as least as it pertained to strawberries or cantaloupe. To know this presupposes a distribution system that makes these diverse human-strawberry/cantaloupe relations possible (how could they have made these distinctions if they only knew what they grew?). But our farming ancestors' bodies were tuned to other subtleties, like a food's age and its variety/breed. Cookbooks from the 1800s, for instance, make repeated reference to the age at which an animal should be slaughtered and the best varieties of vegetables for canning and pickling (see for example Wecox 1885: 202). I know more than a couple of famers that are still in possession of that knowledge, being able to tell with surprising accuracy the age of the animal before it was slaughtered through a visceral combination of visually studying, feeling, and smelling the meat in the uncooked state.

I am hesitant to dismiss this knowledge as something tied closely to class—what Bourdieu (1984: 13) calls "cultural capital". As I've talked about before, taste is to a significant degree socially mediated. The ability to distinguish the subtle differences between wines—in terms of, say, their "notes" and "bouquet"—is often cited for its function as a class marker (Charters 2006: 173; Garson 2007: 55; Guthman 2002: 295, 300). The acquisition of this knowledge requires time (to practice drinking wine), money (to purchase the wine and perhaps sessions at a wine tasting class), and access to particular social networks (if one didn't know anyone with this knowledge it would be considerably more difficult to acquire). Remember also that we are not just talking about the acquisition of knowledge here but also physical routines, such as learning to "properly" swirl the glass of wine so as to release the aromas and open it up, slurping the wine, and swirling it about in one's month before either swallowing or spitting it out. These routines too take time, money, and social networks to develop before the wine tasting performance can be successfully pulled off.

Those interviewed were not rich, according to their own admission. (To recall, all but four respondents located themselves in the category of "middle class".) I spoke with mechanics, nurses, school teachers, beauticians, retired farmers, and convenience store clerks, just to name a few of the occupations possessed by those interviewed. I think an important distinction needs to be made between being able to successfully perform the act of wine tasting and being able to discern between, say, a Fragaria virginiana strawberry (also known as the Virginia strawberry) and a Diamante strawberry. The mark of distinction comes from having internalized "proper" codes and practices. There are specific terms and actions that must be displayed if one is to show their status—their cultural capital—when tasting and drinking wine. I did not come across these socio-embodied markers in respondents' knowing of "fresh" food. There was no universal language that would suggest a "proper" way to talk about strawberries. Likewise, there was no one way to successfully pull off the performance that went along with accessing the freshness of a cucumber or the ripeness of a cantaloupe. To be sure, people had their own thoughts about how best to go about this doing this. In the case of judging the ripeness of cantaloupe, some liked to smell, others liked to shake, still others preferred the method of thumping the fruit with their knuckles. But, as best I could tell, whether one thumped or sniffed or shook was not "read" by others as representing a social cue about their status. In the words of one respondent: "Being color blind I had to come up with different ways to judge this stuff [fruits and vegetables]. But I don't think that's all that different from others' experiences. […] Through trial and error we all learn our own ways to judge the quality and freshness of food. What works best for me might not work for you. It's not a universal".

This ability to fine tune one's taste buds runs counter to the taste regime of industrial food, where distinctions as those discussed above are harder to make (distinctions are still made, they're just different). This is in part due to our bodies having been tuned to food that is mass-produced and standardized. A common critique of Global Food is that it "flattens the sense of variety" creating a "virtually uniform meal experience" (Miele and Murdoch 2002: 314; see also Warde and Martens 1999: 30). Respondents talked about how things like CSA, farmers markets, and backyard gardens work to preserve these disappearing—or flattening—experiences and sensations.

When I talked to Jeff it was his first year with this particular CSA. Having recently moved from Montana he was still getting used to what he referred to as the "bounty of Iowa". Jeff's palate was heavily conditioned by what he called "road food". He was a semi-truck driver for eight years before moving to Iowa. But his understanding of food was changing.

> "We grew up eating what I'd now call crap food, fast food, frozen dinners, boxed 'baked' [he lifted both hands to make scare quotes in the air] goods, soda pop. That's what I lived on for most of my life. I didn't know better. That's really the only way I can explain it. Just didn't know better".

> "So you're telling me it's different now, what you like, don't like?", I asked.

> "Yeah it's changed. Big time. My wife [who he recently married] was the one that got me to change my diet. I didn't want to at first but, after a while, it wasn't a hard sell. [...] What was I thinking eating those shitty Hungry Man frozen dinners with their tasteless vegetables and actually thinking they were good? [chuckles] You couldn't pay me to eat that crap now. But that's just it, I didn't get that type of food when I was younger, growing up. If I did I'm sure I wouldn't have acquired the taste for many of the foods that I did".

This was a theme mentioned repeated during the interviews: that Global Food tunes bodies for the taste and sensations of Global Food. That's what Jeff was getting at when he chalked up liking Hungry Man dinners to years of exposure to this food. And how, having acquired the tastes of non-Global Food, he now thinks that much of that food is (in his words) "crap". Jeff's tastes are growing out of tune for Global Food.

Echoing Jeff, Bill told me his family "eats fewer vegetables in the winter because frozen and canned just don't taste the same [as fresh], not by a long shot". When asked if his—or his family's—diet has changed in other ways since becoming a CSA member he offered the following thoughtful response:

> I wouldn't say that our diets have undergone a revolutionary shift by any means. We're not vegetarian or vegan or hung up on only eating organic. We're not even all that big into eating local. I mean we certainly try in the summer but in the winter months there are things we like to eat and eating local just isn't going to cut it. But we're a lot more aware now. Yeah, that's a good way of putting it. We're more aware since joining the CSA. We're more careful, or maybe the word is more selective, about what we buy and eat. We take more time at the

> grocery store, for instance. We just don't grab whatever's up front when we're
> picking vegetables but we'll search for the good ones. And if there are no good
> ones we'll go without, which is why we like farmers markets and CSAs in the
> summer, because we can always find stuff that meets our expectations. [...] As
> a whole, the experience has made us bigger fans of more locally produced food
> because often store bought food just doesn't meet our standards.

This was a theme repeatedly discussed by respondents: when bodies become
tuned to a grip on food through alternative sources, Global Food ceases to be as
attractive.

Nevertheless, I don't want to suggest that Global Food does not tune for
distinction. Those who argue that Global Food "flattens" (Miele and Murdoch
2002: 314) our food experiences are overstating what's going on. I know people
who have become tuned to distinguish between, say, Coke and Pepsi (my wife) or
between Heinz ketchup and the generic variety (my sister) or between Hidden Valley
Ranch dressing and reduced fat Hidden Valley Ranch dressing (my dad). And I can
certainly distinguish between the McDonald French Fry and those from Burger
King (most everyone I know can). My point, however, is that these distinctions do
not threaten the dominant system of food provisioning. When people talk about
two-party electoral politics in the US they often liken it to the "choice" between
Coke and Pepsi—the point being a lot of decisions have already been made before
we can have our say as an electorate about which caffeinated, carbonated cola
drink (or Democratic/Republican candidate) we prefer. Among those interviewed,
the CSA lived experienced seemed to nurture distinctions that went beyond the
question of "Which cola?". It did this by making the question itself irrelevant, by
tuning individuals for tastes not as easily satisfied by Global Food.

Making Absent Farmers More Present

Previously I discussed Heidegger and his insights into the absences and presences
to being. Yet this earlier discussion missed something important. If I may refine
Heidegger's language a little, what he is really talking about are present absences
(the presences that we don't apprehend) and present presences (the presences that
we do apprehend). This leaves still unaddressed the presences of a thing that are
not there, what we could call absent presences.

Scholarship on the subject of absent presences uses the term to convey issues
similar to that found a few sections back when I talked about how readiness-to-
hand and present-to-hand are socially mediated effects. For example, Slocum's
(2008) study of a Minneapolis Farmers' Market highlights how geographies of race
are continually played out within this space as definitions about local consumption
are shaped predominately by individuals from white, middle-class communities.
Similar arguments are found in Guthman's (2008) analysis of food justice projects,
where growing, donating, and educating African American communities about
food production and consumption were found to reflect the "white desires" of the

volunteers more than the needs of the communities they seek to serve. In either case, issues of race and ethnicity, though absent at the discursive level, have a real presence in how these local food systems are shaped—hence the term absent presence (see also Mansvelt 2009).

I would like to use the term a little differently, which would add another fold to our understanding of our grip on food. By absent presences, I'd like to talk about presences that are absent. This gets us back on the subject of thing-in-itself and how, if tuned properly, we can apprehend those presences of the food chain that are absent at the point of consumption. Recognizing absent presences is thus a step in the direction toward what Marx would call the de-fetishization of the commodity. Food labels and certification programs attempt, among other things, to do precisely this: to make absent presences more present. As it turns out, at least according to some of the respondents, the CSAs studied also had the effect of bringing into better view those phenomena of our food system that are not physically present. The absent presences most frequently discussed were famers, who are otherwise invisible to bodies tuned by Global Food.

> Shannon: "You really want to see farmers get their fair share. That's partially what's attractive about CSAs, you're not seeing the processors getting the lion's share of each dollar you spend on food. Farms should be getting that, not some multinational corporation that doesn't even grow their own food. [...] I always thought farming was hard work, but now, having been a member a few years now, I know it. When I see that cheap food at Wal-Mart I often wonder how the farmers of that food can make a living. They can't as far as I can tell".

Another respondent talked specifically about how, thanks in part to her CSA experiences, she can "see" farmers that many of her friends cannot:

> Maggie: "It's easy to forget the people behind the food. Most of my friends can't see them. Not that I'm surprised. I was no different when I was younger. Not thinking about the farmers and pickers working behind the scene is pretty common. If it wasn't, I think we'd be more critical of all this so-called cheap food the system provides us".

Similar sentiments were conveyed by Nicole, who recalled a past experience with a friend's child to make her point:

> "Her kid didn't know that milk was from a cow. I think they thought it was like pop or something; that it was made in a factory or something. [...] My point is that if you don't even know something comes from a farm why would you think about the farmers making it? At the very least, experiences like these [from CSA] show people where our food comes from. It gives them some basic knowledge about food that then hopefully we can build on to get people to begin thinking critically about conventional agriculture".

Note also how the quotes touch on how Global Food *needs* us to forget. For us to perceive the food supplied by this system as "cheap" requires that we remain blind to some of its most important actors: farmers. The understandings nurtured through CSA threaten Global Food's dominance because they bring these actors back into focus. They help make people understand food as more than a thing-in-itself, to include the presences of absent farmers.

Talking about expressions of care in a globalized world, Zygmunt Bauman (1993: 166) argues that "moral concern would reach its highest intensity where knowledge of the other is at its richest and most intimate, and that it would thin out as knowledge tapers off and intimacy is gradually transformed into estrangement". The extent to which Global Food stretches space and time, to the point that it becomes quite easy to forget about its absent presences, threatens this knowledge of the other. As a result, we are witnessing what Bauman (1991: 193) calls the "vanishing point of moral visibility". Global Food does not encourage, in other words, immoral eating, but rather amoral eating. One respondent touched on this very point when they remarked how "eating industrial food doesn't make you a bad person, though I would be willing to bet you don't think much about where you food comes from and how it is made". To put it another way, respondents seem to be suggesting that CSA encourages reflexive ethical reasoning. Singer (1995: 222) describes this reasoning as follows: "From this perspective, we can see that our own sufferings and pleasures are very like the suffering and pleasures of others, and that there is no reason to give less consideration to the sufferings of other, just because they are 'others'". Such reasoning is captured nicely in the following quote from Elena: "That farmer could be me. And if it were I'd want to be paid fairly". Being able to put yourself in the shoes of the "other", whether physically present or not, went a long way in getting people to think more broadly about food. And it made respondents think twice about the "value" and "cheapness" of Global Food.

Thomas Nagel (1970) nicely summaries the problem with getting people to act by moral reasoning alone. The idea that a single robust principle could be used to persuade everyone to act presumes that this consideration, whatever it might be, can be successfully conceived by theorists and then communicated through pedagogic practice (Barnett and Land 2007: 1070). Yet, as I'll detail time and again in this book, ethical motivation is a relational effect, conceived through one's *doing*. And because of this, because we all *do* differently, ethical motivations take all forms. As Nagel (1970: 3) explains, ethical arguments "must rest on empirical assumptions about the influences to which people are susceptible". If these assumptions are not held by those to whom a moral argument is addressed then the argument has "neither validity nor persuasive force" (Nagel 1970: 4).

I mention this because while everyone interviewed admitted to long knowing they should care about the producers of their food—a few were even able to tell me how many cents on every dollar spent on food goes to farmers—this ethical acknowledgment failed to have motivational force. Knowing what's "right", in other words, wasn't enough to get respondents to act in accordance with that moral imperative; to get them to act, in other words, as they *should*. This moral imperative—something like "thou shalt act in ways that care for farmers"—predated respondents' participation in CSA. It was not enough, however, to elicit meaningful behavioral change in the name of those responsible for feeding them. Not until respondents began placing a face on the otherwise nebulous term "farmer" did they start feeling like they *really should* think more about those who raise their food. One respondent put it to me in the following way: "I think most people have an idea that farmers aren't the ones making all the money; that it's the corporations that are getting an unfair share of the profits. I've known this for a long time. […] It wasn't until I started to call some farmer friends that I started to feel like I really should care about whether they're getting their fair share".

Another individual talked about "not really realizing all the work that goes into farming" before becoming a CSA member. And how now, "knowing what growers do, [they] feel guilty buying food with money that doesn't go directly into their pockets". This unawareness of the work that goes into raising and producing food has been noted by others (see e.g., Freidberg 2009: 7) as helping to stifle social (and labor) reforms because it further fetishize the commodity by making invisible the bodies doing the work. In CSA the spatial distance is not as great, making it harder to hide the fact that some*body* is responsible for growing the food. This, in turn, appears to have made respondents feel more like they really should care about those growing their food and, in many cases at least, act accordingly.

"Quality" as Embedded Process

"They're ready!"

Faye was standing over a cucumber mound holding something about three inches in length. It was a cucumber. Faye makes her own pickles and is very particular about just when the cucumbers need to be picked for this process. Walking towards Faye I asked, referring to the cucumbers, "How do you know when they're ready to be picked?"

"Oh, I don't like them too big. They're better small, when they're still firm. See here". Holding out the small three inch cucumber she took its ends between her fingers and twisted. "See how little it gives. That makes for a crunchy pickle". Taking the cucumber firmly in one hand she continued, "You can also tell its ready by just squeezing it, like this".

"So what are you feeling for?"

"Firmness. It's hard to put into words. I've been doing this a long time. You just know, I guess".

Let us think about Faye's response to my question in terms of tacit knowledge, a term introduced in Chapter 2. Michael Polanyi (1966: 4) famously quipped "we know more than we can tell". Polanyi point was to highlight how *all* knowledge has a component that cannot be codified, the tacit *dimension* of knowledge. Faye cannot entirely explain what makes a cucumber ready for pickling. This is not, however, due to a linguistic deficiency on her part. Faye simply, to use Polanyi's words, knows more than she can tell on the subject of picking the ideal cucumber. For her, the knowledge required to pick cucumbers can only be acquired through the act of picking cucumbers, pickling what you've picked, eating the pickles, and going from there.

Bourdieu's (1995: 248–50) distinction between the *modus operandi* and the *opus operatum* is helpful here. The former speaks to an understanding of a task from the point of view of someone practically engaged within it, whereas the latter refers to understanding abstracted from practice after the fact. To help convey his point Bourdieu likens *modus operandi* to the act of traveling and the countless decisions the traveler is constantly confronted with that inevitably shape the length, direction, and experience of the trip. Conversely, the *opus operatum* is likened to recreating the trip on a map. While useful the latter position gives very little guidance into the complexities of practice for it provides no feel for the trip as a lived experience. In fact, the *opus operatum* can actually mask more than it illuminates, particularly when the "journey" (e.g., picking the ideal cucumber) is heavily imbued with tacit, lived knowledge. Respondents frequently conveyed having this more-than-they-can-tell knowledge that is acquired through lived experience; an experience that the CSAs studied helped to nurture and reinforce. This was particularly the case when it came to descriptions of "fresh" and "quality".

Take the following comment made by Joe, a recent CSA member: "Getting food from here taught me the difference between grocery-store-fresh and garden-fresh". There is a lot packed within this statement. I asked him to explain. "I don't know", was his reply. "There's just a different standard when you go to the store. You look for different things, sometimes because you have to. Corn that's still in its husk and corn in a can are apples and oranges [meaning they are qualitatively different]".

Think of the standards you employ when purchasing canned food (assuming you have made such a purchase). Rarely is your attention drawn to the food itself. Rather, you look at the can—is it dented?—or the purchase-by-date stamped on the container's bottom. The food, what we actually consume, possesses a readiness-to-hand quality. It is largely transparent. "Fresh" within Global Food possesses a significant representational quality to it; much of it has been reduced to purchase-by-dates or heuristics like "do not purchase dented cans". Freshness as it pertains to food from CSA, however, cannot be unproblematically reduced to words or maxims. It has a more-than-one-can-tell quality to it, which can only be learned by physically encountering (manipulating, smelling, feeling, tasting, shaking …) it. We still know Global Food like a body. But within this system our bodies are tuned differently, such as to the feel of dented tin, the sound of air escaping from an air-sealed bag, and the sight of torn plastic around our meat. Joe was onto something when he distinguished between grocery-store-fresh and garden-fresh.

For a point of contrast, recall from earlier the understandings of freshness held in nineteenth-century cookbooks. Our bodies have become tuned to food differently following changes in our lived experiences, which in this case creates different understandings of what constitutes fresh.

The last decade has witnessed what has been called the "quality turn" in agro-food studies (Goodman 2003: 1; see also, e.g., DuPuis and Goodman 2005; Mansfield 2003). As Mansfield (2003: 11–12) reminds us, "Quality is neither a subjective judgment (what different people like), nor an objective measure (the characteristic of a commodity), but instead is produced within relations of commodity production and consumption". I would like to temper this statement a bit, otherwise it could be read as though definitions of quality emerge out of only structural relations and that the concrete specifics of everyday life play no role in this process. Earlier in that same piece Mansfield (2003: 9) explains how "which characteristics count as quality is defined within the production network". I don't deny this point, but I also think quality is negotiated (and continuously re-negotiated) through embodiments closer to the fork-end of the food system. To not make this point explicit risks painting a picture of consumers as dolts incapable of being able to generate food knowledge for themselves.

Evidence of this meaningful meaning generation can be seen by how grounded definitions of freshness and quality, drawn from CSA experiences, altered respondents' perception of Global Food. As bodies became tuned to definitions of quality and freshness that emerge through CSA-embodiments they likewise became out-of-tune for the sensations of Global Food.

"I wasn't much of a fan of homemade pickles at first". We heard from Jeff earlier, the recent transplant from Montana who spent a good chunk of his life (eight years) on the road as a semi-truck driver. "It took a while. I think I got conditioned to the salt of the store-bought variety. Homemade weren't salty enough for me".

"It sounds like you're saying you have different standards now for evaluating pickles? Am I hearing you correctly?", I asked.

"Yeah, that's fair. It's weird; it's almost not even about the taste now. Or, it is, but a big part about what I look for in a pickle now is its crispness. A good pickle needs to snap in your mouth, you know? [...] I have a hard time finding a store bought variety that can give me that. The snap of a Vlasic pickle just doesn't do if for me anymore. [...] Honestly, now that I look back I don't know how the hell I knew how to pick out a good pickle—what had the flashiest ad campaign? I guess I did just mention Vlasic (chuckles)".

It appears that even the objective state of snappiness, which has long been a gold standard sensual experience linked to pickle quality, is contested terrain. Global Food has its definition, which according to Jeff is epitomized by the Vlasic pickle and the pickle-eating stork named Jovny (a stork because, as the stereotype goes, pregnant women crave pickles). But not all pickle "snaps" are the same. For Jeff at least, his understanding of a quality pickle snap changed as he came into greater contact with homemade pickles. As his body became tuned to homemade pickles, store bought pickles began to feel—literally—increasingly out of tune.

Global Food, while undeniably having a tacit dimension, is understandably dependent upon explicit knowledge. I say "understandably" because explicit knowledge travels well. Tacit knowledge, conversely, does not, which is why it is sometimes labeled by economists as "sticky" (Lang 2001: 46). Try efficiently communicating in words (or pictures, numbers, etc.) how to ride a bike and you will quickly learn why the "sticky" metaphor is apt. As anyone who has attempted to teach another to ride a bike knows, this knowledge only transmits at great expense to both the communicator and the receiver, such in terms of lost time, personal injury, and/or property damage (most notably to the bike).

Advertisements that help drive the consumption of Global Food rely heavily upon knowledge that travels easily. Jeff admitted to how the Vlasic advertisements previously helped him evaluate his pickles, with its pickle-eating-stork displaying the trademark Vlasic "snap". Evaluating the freshness or quality (or snappiness) of food based upon representations gives an advantage to Global Food because that is a form of knowledge that it can disseminate with considerable efficiency thanks to the deep pockets of agro-food processors. Yet these dynamics are altered considerably when definitions of quality, freshness, and snappiness become something more experienced than explained. This change gives an advantage to spaces like CSA.

Just so there's no misunderstanding: Global Food does deal in experiential freshness. Yet this is a "mass-produced, nationally distributed, and constantly refrigerated" freshness (Friedberg 2009: 2). Freidberg (2009: 2) calls it "industrial freshness". Industrial freshness is a freshness that travels well; freshness *because of* transportation and processing rather than despite it. This "freshness", for example, requires fruit that can withstand mechanical harvesting and the brutalities of transportation, that ripen off the vine during transport, and that still deliver, as in the case of Red Delicious apples, the all important "crunch" to consumers when they bit into it. Then there is the "fresh cut" version of the industrial apple, which involves a calcium citrate formula that keeps the fruit looking "fresh" for up to 28 days (Friedberg 2009: 156).

Not all types and varieties of fruits and vegetables are amendable to the requirements of capital (see Mann and Dickenson [1978] on barriers to capitalism in agriculture). Some foods have proven easier to appropriate than others. This explains, for instance, why you won't find Summer Rambo apples or Amish Paste tomatoes at your nearest Wal-Mart. Industrial freshness is a homogenized freshness, where what you see, taste, feel, and smell varies little by store, made possible by products that "fit" the needs of capital (or they are substitutes for commodities less amendable to these "needs"). Those products that populate our grocery store aisles are there, more often than not, because they have proven amendable—perhaps you might even say they've become "tuned"—to the harvesting, transportation and processing demands of large scale, global food production.

I am in no position to claim that there is anything inherently wrong with the experiences associated with industrial freshness. My point, rather, is that it's an understanding of freshness that, while a challenge to agro-food processors, does not threaten the system they are wed to. I am sure it would be easier and more profitable for agro-food companies if they could sell "shiny red apples stuffed with cotton" (Freidberg 2009: 155), like what they sold consumers a century ago. Realizing how unlikely it was that the general public could be tuned to choose those early industrial "apple" forms, Global Food made the necessary adjustments (capitalism can be remarkably resilient). Now we can find at any grocery store apples that not only crunch when bitten into but that are also remarkably sweet.

Conversely, understandings of "freshness" nurtured through the CSAs studied were largely defined irrespective of the requirements of capital. Respondents were therefore exposed to sensations—tastes, colors, odors, textures, and even "crunches"—not often provided by the fruits and vegetables that now colonize our grocery store. And as respondents grew used to these sensations industrial freshness slowly became, well, not so fresh.

Visual Distinctions

"That's just extra protein!"

That was the response I received from a CSA customer when asked if the borers (small worm-like creatures) on some of the ears of corn bothered them. There is nothing particularly novel in this response. I've heard it before. Besides, the borers are usually not eaten. Even this respondent admitted to cleaning her corn of borers and bugs during preparation. Also, boiling the corn in water—a popular method of cooking corn-on-the-cob—virtually ensures that any creatures are killed (and cooked). But there is still something of sociological significance in this statement.

I know a lot of people who would refuse to eat corn that knowingly once harbored borers (and who certainly would never eat corn that still might contain a cooked borer or two). Having been around sweetcorn patches my whole life, I know how common the borer-cob relation is (particularly when one chooses not to use pesticides). By participating in a CSA, this relation became apparent to more than one respondent.

"I would have never even thought about buying corn that had worms a couple years ago". Nora, at the time of the interview, was a recent CSA member. Pointing to a small corn borer on a recently shucked ear of corn she continued, "Back then, you would never have gotten me to believe that corn with this little guy on it is perfectly fine to eat. Never".

"What changed?", I asked.

"Well, coming here taught me just how normal an ear of corn with a worm on it is. Sure, I might have been eating worm free corn all those years but that was probably only made possible with pesticides. [...] And I just needed to work through the term 'worm' itself. Once you see one and let it run across your hand, with its big fat mushy body, you realize there's nothing to be afraid of".

As Nora learned, the borer-free field of sweetcorn is in large part a product of conventional farming methods and repeat pesticide applications. But it also hints at a broader consequence of the industrialization and globalization of food. Namely, it gets us into talking about the type of aesthetic that this system helped bring forth. For example, the ubiquitous Cavendish cultivar (yellow and between 15 and 25 cm in length) has been setting the aesthetic standard for our understanding of "banana" for over 50 years (just like the Gros Michel—aka "Big Mike"—cultivar did until the 1950s before it was discovered to be highly susceptible to Panama disease). As Michael Pollan (2001) describes in his book *Botany of Desire*, the apple aesthetic too has become constrained over the course of the past 200 years. The state of being for the apple has become increasingly red, increasingly worm-free, and increasingly sweet.

Stepping outside the conventional system of food production revealed an array of sensations and experiences that caused at least some respondents to reevaluate their sensory priorities when evaluating food quality. One individual explained it to me as follows: "When you're at the store it's all about how something looks. You've got meat that is injected with coloring and the vegetables are made to look perfect, no blemishes, nothing. But here you learn to work with more than how something looks".

The aesthetic of Global Food relies heavily on sight (Hutchings 1977: 267–8); a reality helped along by our detached relationship to its systems of production and provision. When our food comes encased in plastic, frozen in a bag, or sealed in a can we tune our bodies to accommodate this new medium. But on the CSA farms, participants nurtured an aesthetic that was more sensual than that demanded at the grocery store, where consumers depend heavily upon visual cues. At the CSAs studied, a "quality" ear of corn or head of cabbage or tomato could very well not make the cut of what is considerable acceptable for a grocery store because of its appearance. In the words of one individual, "that is exactly why the world needs more CSAs—so more people can learn that good food shouldn't be equated with looking 'just right' if that means it has to be doused with chemicals".

Earlier I spoke of how Global Food tunes bodies to know food at a level that is more than skin deep, which involves experiences that go beyond the visual. This is not meant to diminish the importance of the visual for Global Food. Susanne Freidberg (2009), in her captivating book *Fresh: A Perishable History*, describes at length the importance of the visual for food companies when it came to getting people to buy their products. Fruit growers associations have sponsored contests among store owners to see who could come up with the most pleasing fruit display (pp. 142–3). Visually rich advertising campaigns were used at the turn of the last century to convince consumers of the quality of the industrial food products (pp. 140–41). Some pear growers in the Montreuil region of France in the late 1800s even went as far as to wrap each immature fruit in a paper bag while it still hung from the tree. Though immensely labor intensive, this method protected the fruit from weather damage and the snout moth. But most importantly it created a pear with extra brilliant hues once the bag was removed two or three weeks before harvest (the bag condition increased the fruit's photosensitivity) (p. 148). This technique helped visually separate the pears from this region from other's in Europe, allowing Montreuil growers to fetch a premium price for their labors.

I recognize that humans have long found visually captivating foods desirable. The beautiful, whether talking about clothing, horses, or food, had long been a powerful mark of social distinction in societies (Adorno 1984). I am not arguing that a visual aesthetic is something new to our knowing of food or that this aesthetic is the child of Global Food. It is clear we have long relied upon the visual when assessing food (Lieberman 2006). My concern lies in the *degree* to which we've come to rely upon this visual aesthetic. Since our first impression of a commodity is often visual I would expect food companies to be devoting a tremendous amount of effort, money, and time to making their products as optically captivating as possible. How else, when the "choice" is between Coke and Pepsi (or between two brands of kidney beans, nacho chips, frozen pizza, red apples, etc.), are they to distinguish their artifacts from others on the market? Dyes and colorants, calcium citrate (from the aforementioned "fresh-cut" fruit example), mineral oil sprays (which coat industrial eggs to replace the natural albumen bloom and prevent them from breathing and going stale), and pesticides (e.g., to keep our corn borer-free): these are tools utilized by Global Food to bring visual beauty to the masses. For it is often not until certain visual requirements are met that the other senses are given a "say" in the matter. Respondents acknowledged this emphasis upon the visual, especially before their involvement in CSA. Charles was especially clear on this point, noting:

> I was much pickier about how something looked before joining [the CSA]. If it didn't look a certain way I wouldn't give it a second look. Never even gave it a chance. […] I'm not as likely now to let the look of something get in the way of giving it a chance. I learned long ago that just because something looks good doesn't mean it will taste good but for some reason I still relied heavily upon those visual cues. Maybe it's because I know more about this food, where it's coming from and who growing it, that I feel comfortable paying less attention to superficial things like bruising and discoloration.

Clearly we know Global Food in ways other than through sight. But I think there is sufficient empirical evidence to argue that we've gotten a bit carried away in our attention to the visual when it comes to judging Global Food. I am thinking particularly about the tremendous *waste* of this system, much of which is undeniably because of this visual aesthetic. For instance, I recently came across an article from the *Brisbane Times* (Australia). It contained the following rather startling fact: in the state of Queensland alone over *100,000 tons* of bananas are disposed of each year for not meeting cosmetic retail standards (Hurst 2010). We can debate about whether sight represents a good judge of a food's taste or nutritional character. But one cannot dispute that Global Food takes the visual aesthetic to a fundamentally unsustainable level (see Stuart 2009).

While on the subject of distinction, I recognize that others have found spaces like CSA to be equally guilty of perpetuating status (or race, gender, class) based distinctions (see e.g., Guthman 2003, 2008; Slocum 2008). I found no evidence that such taste-based discriminatory judgments were being nurtured within the two CSAs studied. Since CSA customers essentially received *the same* foods I do not believe distinction based upon visual cues carried much cultural capital within the networks observed. Rather, the real distinctions resided in being able to sensually separate industrial food from that obtained from an individual's local CSA. Tuning bodies to look beyond the visual gave an edge, at least according to respondents, to less processed foods, like those obtained through CSA shares. Some might argue that this in itself is evidence of social distinction simply because greater social status is associated with whole foods and those who eat these (often more expensive) foods. I don't disagree with this argument but it is also a bit beside the point, at least as it pertains to this analysis. I fear arguments that seek to deny the realness of our lived experiences with food, out of concern that someone might mistake such arguments as essentializing taste, are missing an important opportunity: namely, to understand the food system at the point where the rubber of theorization meets the road of everyday life.

Uncertainty and Risk

Embedded within the previous discussion were concerns expressed by respondents about the level of chemical inputs assumed to be used in conventional agriculture. Many of the individuals I talked with mentioned being, to quote one participant, "worried about the amount of chemicals being put into our bodies when we eat conventionally grown and highly processed food". It wasn't that the people I interviewed knew with certainty that Global Food was, to use the phrasing of an earlier quoted respondent, "doused with chemicals", but they certainly worried that it was. Nick perhaps put it best when he explained:

> There are a lot of reasons why we, why I, join things like CSAs. I think on most peoples' list you'll find something like "peace of mind". While I can't ever be 100 percent sure that my food's safe, there's always the chance of contamination somewhere, I'm pretty confident that what I'm getting here won't get me or my family sick.

There are tremendous uncertainties embedded within the global food system. At the site of production, producers deal with such unknowns as the weather, the commodity market, animal behaviors, and the ever-changing price of inputs. And in response farmers do what they can to reduce these uncertainties, such as buy crop insurance, irrigate, and secure contracts. For consumers, a tremendous amount of work (and money) has been devoted to reducing uncertainties, or at least reducing our perception of uncertainties. This very point, for example, is at the heart of the McDonaldization of society thesis (see Ritzer 2008). Through processes of standardization we can now predict with amazing accuracy, for example, what a banana at the grocery store will look and taste like before stepping into the building. We can likewise be assured that the contents of a can of Hunt's Diced Tomatoes purchased in Juno, Alaska, will taste, smell, and look just like the contents of a can of Hunt's Diced Tomatoes bought in Miami, Florida. Similarly, we no longer have to guess what sort of produce our grocery stores will be carrying on any given day. Whereas having to know what was "in season" was stock knowledge for people a couple generations ago (or at least that's what my grandmother and others from her generation tell me) such information is becoming increasingly less important.

As sociologists of knowledge tell us, the phenomena of certainty and uncertainly are not objectively given (Campbell 1985: 429–34; Shackley et al. 1998: 160–71). For example, in their study of the scientific debate that surrounds the Steller Sea Lion off the coast of Alaska, Mansfield and Haas (2006: 81) note how uncertainty cannot be defined as an objective state alone but is a product of broader social relations: "People produce uncertainty when they agree on what is not known […], but also when they disagree on what is known […], or even when they agree on what is known but interpret that knowledge in different ways". With this point in mind, the authors then go on to demonstrate how powerful interests were able to shift the scale at which the Steller Sea Lion debate was framed to introduce further uncertainty into scientific models. Specifically, the fishing industry successfully framed the scale of the debate to the level of macro-regional drivers, such as climate change, versus the National Marine Fisheries Service, who wanted to focus on local interactions between individual fisheries and the Steller Sea Lion. And, not surprisingly, as "the system" studied expanded in its scope, the scientific models employed rested upon an expanding degree of assumptions and unknowns.

From the standpoint of the consumer, Global Food has been quite successful at either eliminating or masking uncertainties. The size, taste, smell, and appearance of a banana or the contents of a can of Hunt's Diced Tomatoes are almost assured regardless of where in the US you do your shopping. Similarly, one never (or rarely ever) wonders if the local Super Wal-Mart will have exotic fruit like kiwi in stock. Yet, as respondents made clear, Global Food has not eliminated uncertainties. In fact, at least among those interviewed, what Global Food has done is substitute acceptable uncertainties for unacceptable ones. Again, Nick:

> There are things about CSAs I don't like. Nothing's perfect. I don't like not knowing, or at least not being able to plan for, what I'll be getting. I may really want lettuce but if for whatever reasons the lettuce doesn't make it I'm not getting anything. That I don't like. But I can live with that. I can't live, however, with not knowing the growing practices of food that may not just come from hundreds of miles away but that might be coming from another country.

This corroborates a finding made earlier by Claire Larmine (2005) in a study of three different long-term subscription schemes that provide a weekly box of fruits and vegetable to subscribers. According to Larmine (2005: 325), participants saw the schemes as a way to exchange unacceptable uncertainties (such as the use of chemical inputs) for guarantees (such as no chemical inputs) or acceptable uncertainties (uncertainty about the assortment of fruits and vegetables in each box). Global Food does not want either the presence or absence of chemicals to be an issue for consumers. When it comes to food, many today are tuned to things like price, expiration and sell-by dates, and whatever visual cues one can glean through the plastic wrap. Yet what if we suddenly became tuned to the question of whether chemicals were used to grow our food? In such a scenario, it is pretty safe to say that conventionally grown food would find itself in trouble.

> It's not that the absence of chemicals is a major selling point for [CSA] members. The very fact that it's talked about here makes it more of an issue than with conventional food. The topic of chemicals is brushed under the table by Big Ag. They don't even want us talking about that. Here, we talk about it.

For Sara, the source of the above quote, the fact that "chemicals" are part of the discussion brings into view an aspect of Global Food that those associated with this system would rather us not have. Neither CSA studied used chemicals. It is hard to discern whether members' views about chemical inputs representing an unacceptable uncertainty were present before joining the CSA or it if they emerged (or were strengthened) through the CSA experiences. Probably, like Faye, it was a little bit of both.

> Faye: "I was more casual about it before becoming involved [in this CSA]. I ate organic when I thought about it. I ate it in spurts. I mean, I knew about the chemicals and hormones and stuff that are used in conventional food but I didn't always act in ways to lower the risks".
>
> "So you're more consistent now?", I asked.
>
> "I'm still far from being a full-fledged organic eater. But I think I'm more conscious of it now. I think about it more now".
>
> "What's causing you to think about it more now?"

"I wouldn't say it's because of anything specific. [Approximately five second pause.] It's the whole CSA thing. It's just that when you think more about your food, when you know more about it, you start thinking about a lot of things, like whether we should be eating food that's been grown with chemicals".

This is not to suggest, however, that non-CSA customers are somehow deficient in their knowledge about the use of chemicals in industrial agriculture. As others have argued, sometimes what looks to be ignorance is in fact a veiled attempt to preserve a sense of agency (Giddens 1990: 90; Miller 2001: 15). Sociologist Brian Wynne (1995: 380), who is arguably the most famous critic of the "deficit model", once put it this way: "[Ignorance is not a product of] a cognitive vacuum, or a deficit by default of knowledge, but an active construct, and one with cognitive content, about the social dimensions of space". Continuing, he noted that ignorance is "part and parcel of the dynamic of social identity" (Wynne 1995: 380). If we are forced to confront all of the risks that we encounter in our routine lives we would become paralyzed. A level of ignorance is therefore needed to be deliberately maintained toward at least some facets of our existence. For many people, because when pressed most admit to knowing that chemicals, hormones, and antibiotics are used to produce our food (Tucker et al. 2006), this ignorance is directed towards what we eat (Tulloch and Lupton 2002). This explains why "[t]he way people talk about food does not necessarily match the way that they consume it" (Eden et al. 2008: 1054).

Those I talked to for this research, however, choose not to ignore these realities. Faye used to actively forget, as indicated in her statement that she "knew about the chemicals and hormones and stuff that are used in conventional food but" that didn't always cause her to "act in ways to lower the risks". Being part of a CSA served as a continual reminder of those realities, making the act of forgetting all the more difficult. Sara made a similar point a couple pages back when she noted how "the topic of chemicals is brushed under the table by Big Ag". At her CSA, conversely, the subject is repeatedly discussed.

In addition to generating alternative knowledges and meanings it is important that we understand how these spaces also make other knowledges more difficult to ignore. When attention focuses entirely on the "pedagogical" (Hudson and Hudson 2003: 427) aspects of a space the analyst risks casting individuals in the problematic dichotomy of being either "knowledgeable" or "knowledge deficient" (and thus in need of some enlightening). Yet, as the above discussion points out, sometimes non-knowledge is deliberately maintained. CSA appears to make at least some of this non-knowledge, particularly that pertaining to the global food system, more difficult to maintain.

Reflexive Theorizing: The Food-Body Assemblage

Michael Goodman (2010b: in press) recently wrote: "You are what you eat—literally—but also *how*, *when*, *where* and *why* you eat" (see also Longhurst et al.

2009: 333). There is general agreement among agro-food scholars and cultural theorists that food, in a very real sense, makes us who we are. "[F]ood blurs the boundary between internal/external, self/Other" (Gething 2010: 269), we are told. An anti-essentialist spirit seems to inform much of the scholarly food literature. Yet if we look a little closer we find these arguments still haunted by the rock hard categories of modernity (to use Latour's [1993] terminology). Take the following quote. Coming from a work I would classify as embodied food scholarship, the author's intent is to blur boundaries. What is actually accomplished, however, is far less radical:

> Food involves the proximate senses of taste and touch because food is incorporated into the body through the act of eating. The mobility of food into the body involves its passage through the mouth and through the body. The mouth is a transitory space in the relationship between food, self, and culture. The mouth has a "particular role as a sensory body opening" (Falk, 1994, p. 14). The mouth is thus a gateway into the body, a passage from the outside to the inside (Gibson 2007: 5–6).

In a similar spirit, Deborah Lupton (1996: 16–17) writes: "Food is a liminal substance; it stands as a bridging substance between nature and culture, the human and the natural, the outside and the inside".

There is a lot of talk about blurring boundaries in the agro-food literature. Closer inspection yields, however, little evidence that scholars have really moved that far beyond the essentialized categories they claim to make transparent. The inside, the outside, the human, and the natural: these ontologically independent worlds still exist in many so-called anti-essentialist analyses. For in the end what they are talking about is the crossing—through eating—from one world into another.

This is more than just a philosophical problem for me. Essentializing these categories has profound empirical consequences. This is a clear example of why philosophical assumptions matter, even for social scientists; for assumptions ultimately shape what we as researchers "see". If we take the categories of "the human" and "the natural" as given we'll only look at the most obvious place where those two worlds come into contact with each other, which explains why so much theoretical attention has been given to the act of eating. Yet what if this particular methodological stance was itself an artifact of a particular embodiment? The industrial food body often only comes into contact with food near the "end" of the chain, through purchasing and consuming these artifacts. It therefore makes sense that when theorizing about when food "enters" into the body significant analytic attention is placed on the act of mastication. The act of mastication, after all, is perhaps the most visceral encounter most today have with food. Yet what of those bodies that know food in ways that go beyond the act of consumption? What do the categories "body" and "food" look like in those instances?

"What does the concept of 'food' mean to you?" I was talking to Jack, a teacher and a longtime CSA member.

"I could give you the two second definition by just telling you 'it's whatever I eat' but I know that won't do; though, thinking about it, that's a pretty short-sighted definition. I mean, I could get all metaphorical on you and say food nourishes me but not just in those ways we typically think of when we think about food being nourishing".

Intrigued by this answer, and therefore wanting to hear further elaboration, I asked, "So how does food nourish you, exactly?"

"Well, it nurtures me at the table, when I'm eating the food, but also well before that, like when I'm out here [at the CSA] or in my garden. For me, food isn't just what you eat but it also represents a way of life, a life that involves eating as close to the ground as possible [this individual repeatedly spoke of 'close to the ground', which for him meant eating local, whole foods]. I'm consuming food every time I'm in a garden".

As opposed to conventional food (and body) theorizing, which tends to locate the transgression between food and body in the act of eating, respondents spoke of how this transitory relationship occurs much "earlier" and involves an *opened body* rather than a *body opening*. While impossible to trace it back directly to the embodiments made available by their participation in CSA, interviewees repeatedly hinted at how their lived experience with food shaped their understanding of these otherwise pre-given categories. The following are a few representative quotes describing these fluid categorizations and their links to lived experiences like those offered at the CSAs studied.

Nora: "A healthy agro-ecosystem makes for healthy people. I really believe that and this conviction has only been strengthened since I've become involved with [the CSA]. [...] So I guess you'd say you're not just what you eat. How you grow your food shapes who are as well".

Joe: "I eat every time I step foot onto these [CSA] gardens. I'm taking it all in, the sounds, the smells, the friendships made. [...] But unlike when I eat at the table I never really get full out here".

When it came to their understandings of food and body, respondents appeared a little less modern (again *a la* Latour) than those food theorists quoted earlier. While links between the CSA lived experience and a blurred understanding of food and body are suggestive at best this subsection provides at the very least a reminder to theorists to never stop being reflexive. And this goes for something as seemingly obvious as when "food" enters into the "body"; for as respondents pointed out, even the simple act of "eating" rests upon a series of assumptions that appear to have roots in the lived experience. So I agree with Goodman (2010b: in press). We are most definitely what we eat, as well as how, when, where, and why we eat. And the same holds with food: what it "*is*" can be said to be thoroughly entangled in how we relate to it (which, for many embodiment scholars, appears to be at the end of a fork).

Chapter 4
Thinking with Heritage Seed Banks

The idea that our lived experiences with food only shape our understandings of food is clearly shortsighted. Being with food not only informs our knowledge of what we eat. It also feeds into how we know ourselves, our bodies, as briefly discussed at the end of the last chapter. This chapter looks at other lived experiences with food—namely, those found in a heritage seed bank—showing them too to be of epistemic consequence to our knowing the world.

Knowing Nature

Until the sixteenth century, the natural world was largely understood through the texts of Ancient Greek and Roman thinkers, most famously Aristotle, Theophrastus, Dioscorides, Pliny, and Galen. Prior to the Enlightenment, philosophers believed it unnecessary to go out and study nature itself. Rather, these ancient works were believed to hold all the answers about the natural world (Ogilvie 2006: 96–100). This caused scholars prior to the sixteenth century to disregard the particulars of nature, leading them to focus instead on uncovering universals and essences, particularly those related to the medical virtues of plants (Cook 1996: 92–5). The works of Galen, for example, in focusing on the medicinal essences of certain vegetation, have very little to say when it comes to providing morphological details. Galen in fact explicitly argued against the use of such descriptions (Reeds 1991: 22). Medieval theologians believed the careful observation of natural objects for their own sake to be a vice. Scholastic philosophy during this period argued that the book of nature provided little more than an access point into our own spiritual content. Thus, just as one would not study the letters on the page in a book, neither should they study the particulars of artifacts found in nature (Ogilvie 2006: 103).

This way of knowing the natural world, however, slowly changed with the rise of Enlightenment thinking. It was around this time that personal experience began playing a more pronounced role. In the sixteenth century, for example, Francis Bacon argued that an active examination of nature could not only reveal its hidden truths but could be used to benefit all of humankind. It was within this newly emerging intellectual climate that a handful of well known naturalists undertook a radical project: to assess the accuracy of the classics through observation and subject these long held Truths to empirical scrutiny. Rather than merely "correcting" these texts philologically, as had traditionally been done, naturalists, beginning with Leoniceno (1428–1524), wished to compare them to their own observations and to the observations of their contemporaries.

The reasons for this shift toward observation and description are many. As the experiential world of naturalists expanded, via personal correspondents, travel, and the spread of the various empires, a need arose to more accurately talk about the natural world so that these experiences could be shared, compared, contrasted, and delineated between. The Enlightenment itself also created a shift in aesthetics toward nature. Where previously the beauty of this realm was its divine order, now its splendor was located in knowing the particulars of that order. There was also a growing passion emerging throughout Europe and Britain for collecting (Pearce 1995: 1–20). Unique majestic artifacts such as lions and elephants had long been prized by kings and popes as symbols of their power. By the seventeenth century, however, equal status began being conferred onto those who possessed large collections of artifacts less grand in their scale (such as plants, herbs, and rocks) (Findlen 1996: 60–70). This period also saw the separation of two previously wed intellectual pursuits: that of seeking the medical virtues of plants (becoming the providence of herbalists, apothecaries, and physicians) and that of describing a plant's physical characteristics (thereafter becoming the subject of botany) (Ogilvie 2006: 204).

A move was underway where nature began to be observed directly and where particular aspects of those observations were recorded and shared. For much of the 1600s, the mark of the experienced naturalist was travel in the countryside, to experience the artifacts *in situ* (Ogilvie 2006: 142). Yet, as this appetite for observation grew, naturalists sought out additional ways to know and experience these phenomena. Because they could not be everywhere at once, techniques were developed that allowed for the observation of artifacts from the safety of one's home (which was particularly important during the winter months when plant life was difficult to observe in the wild) (Findlen 1994: 153). Gardens and greenhouses emerged during this time to give scientists the chance to observe the life cycles of plants without having to travel repeatedly over multiple seasons (botanical gardens first emerged in a number of prominent European cities, such as Pisa and Florence, in the 1540s). Herbariums likewise proliferated during the sixteenth century (Ogilvie 2006: 158). These collections of dried plants allowed for year-round observation, regardless of weather and season (no doubt why they were also called "winter gardens"). The technology of woodcutting illustrations and the realistic art that emerged during the Renaissance also proved significant. Through these media, observations could be recreated in—and reduced to—visual form.

However, naturalists were no longer merely describing artifacts of nature. While the shift is subtle, by employing these various technological forms—from gardens, to herbariums, visual illustrations, and verbal descriptions—naturalists were, in turn, altering the type of "nature" that they were experiencing. Some early naturalists appeared sensitive to this point. For example, Carolus Clusius was skeptical that a dried plant on a loose sheet of paper could sufficiently represent what had once been a living organism in the wild (Ogilvie 2006: 185). Paintings and woodcuts presented similar epistemic dilemmas. The scientific realism sought by the naturalist was not the same as the artistic realism that had emerged during the Renaissance. Illustrations of plants, for instance, did not contain shadowing, for fear it might hide important features. The various parts of plants were also often spread open, so as to maximize what could be seen. It was also not unusual to draw a plant containing both flowers and fruits, thus showing it in various parts of the life cycle simultaneously.

These "realist" representations thus began to depict entities that *did not exist* in reality. While a description and illustrations of particulars were sought, such particulars were not of individual plants. The particular that was depicted had to be general to the entire species. This can be witnessed in the verbal descriptions given to plants. Rather than reducing an organism to words, whereby all physical characteristics are described in their entirety (recognizing this to be an impossible feat), descriptions were limited to those aspects that would allow it to be distinguished from other plants. Thus, both visual and verbal representations focused upon particular elements, which together did not represent any individual specimen. Rather, they where meant, when viewed in aggregate, to present a *gestalt* of the species itself so as to allow for efficient delineations to be made between it and other species.

By the eighteenth century, another change was underway. At this time, specimens and descriptions from the far reaches of the European colonies were beginning to enter into the experiential world of naturalists at a rate never before known. The exponentially growing list of names, illustrations, and descriptions began to overwhelm the cognitive capabilities of scientists; a problem that was only exacerbated by the fact that no logical classification system as of yet held these entities together. In order to describe, compare, and delineate between this increasingly unwieldy amalgamation of specimens, more time had to be spent gazing upon proxies, such as gardens, herbaria, illustrations, and verbal descriptions. This also meant that less time was spent traveling to observe these entities first hand in the wild. As a result, a further economization of morphological descriptions was sought. Early naturalists provided not only a list of physical characteristics that allowed delineations to be made between species. They also described, for example, where the plant was observed, the time of the year it flowered and bore fruit, and its color and scent. Such rich descriptions were the result of having observed the plant in the wild at multiple points in its life cycle. These non-morphological, non-standardized, highly sensuous descriptions began to disappear as scientists restricted their gaze to more objective qualities.

Whereas the expertise of early naturalists strongly correlated with the frequency they traveled to observe and record nature in its original setting, later naturalists acquired expertise by remaining at home. For example, the French naturalist Georges Cuvier turned down Napoleon's offer to travel with him to Egypt, believing he could better study natural history in a Paris museum (Ogilvie 2006: 142). Soon the most recognized naturalists became those who let others do the traveling, which allowed them to arrange specimens in such a way that was not possible in the wild. For this to happen, however, "nature" needed to be made stable so that the knowledge it held could travel well.

Previous to this, stability was less of an issue as entire organisms were collected. Yet this knowledge did not travel well given the difficultly of transporting these artifacts over great distances (even if they could be transported, animals, for example, often died either during the trip or shortly upon arrival). Technologies therefore had to be developed to improve the mobility of nature. Bruno Latour (1987: 254–7) refers famously to these technologies as "inscription devices". Through these devices, an unwieldy nature is reduced to mobile proxies—such as seeds, bulbs, dried plants, woodcuts, skeletons, and so forth—that can stand in for that aspect of the lived world under investigation. Inscription devices were not perfected overnight, however. Early naturalists who began to rely more upon dried specimens, for example, frequently complained of their non-stable qualities. They protested about how dried plants would often break apart during transport and how the color of flower petals sometimes changed after being dried (Ogilvie 2006: 170). Likewise, early herbaria were first designed to be mobile. Such technologies soon proved to be too cumbersome as collections grew. Nature was then brought to the herbarium in an attempt to resolve this problem. A parallel of this today would be the conventional gene and seed bank, where nature—though today we prefer the term biological diversity—is reduced to such proxies as seeds, semen and eggs, and coded bits of protein (DNA).

"Biodiversity" is Not Biodiversity

In his book *This is Not a Pipe*, twentieth-century philosopher Michel Foucault (1983) questions the received notion of representation in art by examining the works of the Belgian surrealist painter Réné Magritte. Foucault gives particular attention to Magritte's work *Ceçi n'est pas une pipe* (1926)—*This is Not a Pipe*—from which Foucault derives the title for his book. Foucault discusses at length how Magritte's painting of a pipe, combined with the painted words "This is not a pipe", calls into question a host of issues related to visual representation.

From the standpoint of knowing biodiversity the critique of representationalism is particularly apt. Take the case of zoos. The orthodox view of zoos is that they are places of science, education, and conservation (Hahn 1967: 1–10). A closer look, however, paints a different picture of these spaces. In one study, it was found that "zoo-goers [are] much less knowledgeable about animals than backpackers, hunters, fishermen, and others who claim an interest in animals, and only slightly more knowledgeable than those who claim no interest in animals at all" (Kellert 1979, as cited in Acampora 2005: 73). Years later, the same researcher finds that little has changed: "the typical visitor appears only marginally more appreciative, better informed, or engaged in the natural world following the experience [and] many visitors leave the zoo more convinced than ever of human superiority over the natural world" (Kellert 1997: 99). If zoos are one of a small suite of spaces that provide the majority of people in the West their only contact with living biological diversity (Maddox 1991: 457), and if these spaces have "the potential for changing public attitudes concerning conservation of our natural environment by providing a glimpse [...] of its wonder, beauty and mystery" (Polakowski 1987: 16), then we must take care in understanding just what these spaces are conserving.

Zoos give visitors a very distinct grip on the phenomena they claim to be conserving. They are highly controlled and contrived spaces: diets are artificial (no so-called food chain exists here); mate selection is restricted; fighting between animals is limited; animal feces and odors are "managed"; and so forth. The feeling—literally—we get for such phenomena as biodiversity, nature, and wilderness at zoos is constrained and limited. Looking back at my numerous trips to zoos I recall how highly sanitized the experiences were. Spaces were heavily oriented to the optical experience. Yes, other senses came into play. Yet these experiences were often either background supplements—I remember one instance where the sounds of a rainforest were playing through a sound system (as if all rainforests sound the same!)—or management "mistakes"—I once overheard a visitor complaining about how the zoo keepers needed to do a better job cleaning up after the animals.

Sensually speaking, zoos do give us something to grip onto, even if that something is highly problematic. Other conservation practices look, feel, taste, smell, and sound nothing like zoos. Take a conventional gene bank, such as the National Center for Genetic Resources Preservation (NCGRP). Located in Fort Collins, Colorado (US), the NCGRP's mission is directed at the storage of animal (e.g., semen) and plant genetic materials. At the NCGRP (specifically its storage of plant genetic resources), various approaches to *ex situ* conservation are practiced. Here, "orthodox" seeds (those that can be dried and frozen without a significant loss of viability) are dried to a certain moisture level and frozen, in some cases for decades. The conditions for the storage of such seeds are specific to each variety. Generally, however, seeds are dried to a moisture content below 7 percent (for very long storage periods, the lower the moisture content the better), sealed in a moisture proof container, and stored at a temperature of between -14 and -18 degrees Celsius. The success of the preservation technique depends on the continual monitoring of viability. When viability drops below an acceptable level, regeneration and/or recollection of seed is then conducted. Conversely, for seeds that are unable to withstand drying below a certain moisture level (what are called "recalcitrant seeds"), *in vitro* storage takes place. *In vitro* involves the collection and maintenance of tissue samples in a sterile, pathogen-free environment. This technique is also used for the preservation of plants that do not produce seed—what are called vegetatively propagated plants (such as potatoes and many fruit crops).

Only people with clearance can enter the building (while the parking lot is monitored by closed circuit cameras), and for those given access, action is constrained, rationalized, and standardized. There is therefore very little tolerance within this space for the unexpected, lived experiences of everyday life. In the end, the goal of NCGRP—like most gene banks the world over—is the preservation of genetic material. At NCGRP, nature = genes, biodiversity = genes, resources = genes, and knowledge = genes. Having spoken to employees of this facility—

NCGRP is located on the campus where I work (Colorado State University)—I am repeatedly struck by the narrowness of its conservation project. In this space, there is an utter disregard for the sensual. One researcher working at NCGRP memorably explained their conservation focus to me as follows: "Protein, information, and water—that basically sums it up in terms of what we're interested in protecting".

The Case: Introductions and Analysis

The Seed Savers Exchange (SSE) is located near the town of Decorah, Iowa, in the far northeastern corner of the state. Founded in 1975, SSE is a nonprofit organization that both saves and sells heirloom fruit, vegetable, and flower seeds (though there is no universally agreed upon definition, "heirloom" generally refers to varieties that are capable of being pollen-fertilized and whose existence pre-dates industrial agriculture). On this 890-acre farm, which goes by the name of the "Heritage Farm", one will find a diverse genetic legacy. There are 24,000 rare vegetable varieties, including about 4,000 traditional varieties from Eastern Europe and Russia. In addition, SSE possesses approximately 700 pre-1900 varieties of apples, which represents nearly every remaining pre-1900 variety left in existence (out of about 8,000 that once existed). Beyond that, they have approximately 80 Ancient White Park Cattle (only about 800 of these animals remain in the world).

Thus, the practices of SSE go far beyond that of a seed repository. For example, SSE sells its seeds (on site, through a catalogue, and via the internet). They also educate others—through, for example, books, workshops, and lectures—on how to save their own seeds. The "bank" metaphor is therefore problematic as it applies to SSE. Like banks, there is a form of exchange taking place, in that heirloom variety seeds are given to SSE to be propagated (like a form of interest). Unlike banks, however, SSE educates people on how to save their own seeds. While the bank metaphor implies a space that is centralized with fixed boundaries, SSE is best understood as a decentralized space with indeterminable boundaries.

Though SSE is without an explicit mission statement, one could glean from their website a statement that comes quite close to summarizing their philosophy and purpose:

> Seed Savers Exchange is a nonprofit organization that saves and shares the heirloom seeds of our garden heritage, forming a living legacy that can be passed down through generations. [...] Today, the 890-acre Heritage Farm, Decorah, Iowa, is our home—and Seed Savers Exchange is the largest non-governmental seed bank in the United States. We permanently maintain more than 25,000 endangered vegetable varieties, most having been brought to North America by members' ancestors who immigrated from Europe, the Middle East, Asia and other parts of the world. Unlike Fort Knox, Heritage Farm is not surrounded by security fences and guards. Our perimeter is patrolled by Bald Eagles, red-tailed

hawks, deer, raccoons and other wildlife (http://www.seedsavers.org/Content.
aspx?src=aboutus.htm).

Data collection for this study began in the summer of 2004. Twenty-two interviews
were conducted between June and July. Seven non-consecutive days were spent
at SSE. During each visit, I carefully observed the spatial arrangements of the site
and how people negotiated that arrangement. I also randomly approached visitors
and asked if I could interview them as they walked from garden to garden. This
not only allowed me to interview them in the traditional sense of the term but
it also gave me the opportunity to engage in a sort of participant observation,
whereby I participated with them as they made sense of the space around them.
Staff members were also interviewed to educate myself about SSE and to get a
sense of how the operators of SSE understood their efforts. During this phase of
data collection, individuals were asked questions related to: what they understood
the mission of SSE to be; their specific experiences with SSE; what was being
conserved at SSE; their own experiences with saving seeds; and the experiences
they had with others involved with SSE.

After spending several months examining the data for emergent themes I
re-entered the field in the spring of 2005 to explore these themes further. Two
additional visits were made to SSE during this time. During each, interviews
were conducted in the same manner as those the previous summer, for a total of
six additional interviews. In all, 28 interviews were conducted: six staff and 22
visitors. The length of each interview spanned from 20 minutes to slightly over an
hour. On average, interviews lasted between 30 and 40 minutes. As in the previous
chapter, the names of respondents have been changed to protect their identities.

It is difficult to grasp the possibility that our understandings of biodiversity
are missing something. The narrative that science is the accumulation of factual,
objective knowledge is seductive. What could possibility be missing? "We know
more today", I can hear someone saying, "about the world than we did yesterday
and tomorrow we'll know more than today!" But what we know so well is
knowledge that travels well: explicit knowledge—knowledge that can be reduced
to words, pictures, a figure point, and binary code. Knowledge too exuberant to
be confined within representations has lost its currency, no longer existing on
an equal plane with objective knowledge. Alfred North Whitehead (1938: 211)
explained this separation of experience as follows: "[Science] only deals with
half of the evidence provide by human experience. It divides the seamless coat—
or, to change the metaphor into a happier form, it examines the coat, which is
superficial, and neglects the body which is fundamental". Anthropologist Clifford
Geertz (1996: 262) made a similar remark when we noted that "no one lives in the
world in general", though scientific knowledge assumes this to be true.

What we need is contrast; something to bring the background into the
foreground. To understand what conventional conservation practices (where
attention is directed at such artifacts as genes and species) miss requires the
examination of a space like SSE where a distinct lived experience is nurtured.

The analysis that follows is inherently incomplete. I could no more describe the experience felt by respondents while in this space than convey the essence of a pipe through a painting. But I can, through the voices of respondents, give the reader at least a taste of some of the other phenomena that SSE is working to conserve in addition to the objective artifacts of genes, seeds, and organisms. Doing this gives us a glimpse at how biological systems are more than the sum of their parts, and why qualities matter as much as quantities.

On Place, Practice, and Memory

In *Art of Memory* Frances Yates (1966) famously describes the mnemonic technique used by the orators of Ancient Greece and Rome. To recall from a previous chapter, rhapsodes were the primary sources of history and the means by which culture was transferred across generations in pre-alphabetic societies. But because there was no text—and thus no-thing to memorize—classic orators utilized a technique of recall different from that conventionally employed today; a technique regularly practiced by orators in the West until the late Renaissance with the diffusion of typographic text. According to Yates, classic rhapsodes imaged a grand palace. This space was expansive, constituted by elaborate rooms and halls that were each filled with ornate details. The orator would then envision himself (they tended to be male) navigating this space depositing aspects of the story along the way. To recall the story, then, all the person had to do is retrace their steps through the palace. Each room would recall a particular topic and each thing in the room a specific phrase or point to be made.

Before one writes this off as too strange to be true, a growing body of anthropological evidence suggests that this cognitive technique is quite widespread. Yet rather than housing this linguistic–topological field within the structures of private imaginations, this literature points to the lived world itself—what one anthropologists has called the "nourishing terrains" (Rosa 1996) of the mind-body—as the store of knowledge. Anthropologist and musicologist Steven Feld (1996), for example, uses the concepts of "soundscape" and "acoustemology" to speak about mnemonic techniques of the Kaluli, rainforest dwellers in the Southern Highlands province of Papua New Guinea. Feld argues that the sound, feel, and smell of water are recalled through songs that in turn evoke a host of sensations and memories—a literal poetics of place. Describing the intensity of these place-based memories, Feld (1996: 114) writes,

> The aesthetic power and pleasure of Kaluli songs emerges in part through their textual poesis of placename paths. Composed and performed by guests in ritual contexts to evoke tears from their hosts over memories of persons and places left behind, these songs can also be sung during work, leisure, and everyday activities by women and men as they move through and pass time in forest locales. In both ritual and everyday contexts, the songs are always reflective and contemplative, qualities enhanced in each instance by construction of a poetic

cartography whose paradigmatic parallelism of path making and naming reveal how places are laminated to memories, biographies and feelings.

But this is not to imply that one has to be an indigenous forest dweller to find place good to think with. Michael Mayerfeld Bell (1997), in an article titled "The Ghosts of Place", gives a thoughtful and personal account of the mnemonic powers of the world in contemporary society. The house where one grew up. The spot on the river where they went fishing as a child with their grandfather. The hospital room where their best friend died. Places like these, Bell explains, are quite literally event-*full*. Stepping into them one cannot help but think certain thoughts, see ghosts, feel a presence.

"See this; I'm holding a bit of my childhood here". I was standing outside the visitors' center talking with Betty. She was holding a packet of broomcorn seed in her outstretched hand. She went on:

> As a kid I used to go over to grandma's and grandpa's house; they had a farm and planted a little bit of everything. [...] The broomcorn was always my favorite. It was like four times the size as I was so walking between those rows I would pretend I was walking through a forest. [...] But my fondest memory is all the neat things I would make with grandma using that stuff. We did a lot of crafty things with it. [...] We would do that every fall, probably for about 10 years.

Wanting to push her more on one point I asked, "So is that what you mean when you say you're holding a bit of your childhood in your hand—that it reminds you of doing those things with your grandparents?"

> "Yes, it reminds me", she replied, "but it's more than just about memories because the corn actually exists. I can walk through those rows in the late spring and it will be like I'm 10 again. The feeling of the tassels, working the stalk into wreaths just like I did with grandma; it all takes me back".

Broomcorn certainly has ghosts for Betty; but to chalk these memories up to only spirits strips the world of its own eloquent force. I do not think Bell goes quite far enough in his phenomenological account of place. For Bell, place is dead, enlivened only by the ghosts we impart onto an otherwise dead material terrain. In Bell's (1997: 815) words, it is ghosts that "give life to those places". Betty and other participants, however, were clear about things and places having a life of their own. Seeds, for example, have a remarkable mnemonic quality. In addition to having ghosts they have this ability to literally embody the past, present and future. That is what Betty was getting at when she explained how the seeds in her hand were more than just "memories because the corn actually exists". Many of the mnemonic devices of today—like war memorials, paintings, national holidays, and even recorded video of past events—rely heavily upon ghosts to give them meaning. In fact, sometimes the event being remembered is so traumatic that great steps are taken to intellectualize the memory; to distance the body from the past as much as possible. We would never memorialize 9/11, for example, by simulating airplanes crashing into the New York skyline; recreating what could only be described as the indescribable smells, sights, and sounds of the event near "ground zero", in addition to any other sensual experiences that made up this space on that unforgettable day. Rather, we are asked to remember with a moment of silence, of self-reflection, of prayer. We are asked to think about the event; but when doing so please leave the body, as much as possible, out of the experience.

Seeds, conversely, offer no such bodily reprieve. Betty was explicit about this when she talked about the mnemonic powers within the "feeling of the tassels" and "working the stalk into wreaths". Others provided similar accounts whereby the past, present, and future collide within the singularity of the seed. Jon, remarking on the event-*full* nature of the SSE experience, noted:

> I remember coming here for the first time and seeing varieties [of vegetables] that I hadn't seen for ages. I mean, it's one thing to see pictures of Golden Bantam [the first widely marketed variety of sweetcorn introduced by W. Atlee Burpee in 1902], but to come here and see it first hand, to feel it—nothing can compare to that. It's living history.

When asked to explain what he meant by the term "living history" he remarked, "There is something special about the fact that I can grow something in my garden that my dad and his dad grew. I see what they saw, taste what they tasted, smell what they smelled".

Anthropologist Virginia Nazarea (2005: 98–9) wrote about the need for nostalgia when discussing heirloom seed conservation. This is not, however, the type of nostalgia that seeks to arrest change, that longs for the past, and that, on occasion, has racist and xenophobic undertones. The nostalgia Nazarea is referring to draws upon its Greek roots: to return to one's home (verb, *nosto*) and the desiring with burning pain or feeling (noun, *algho*). Unlike conventional understandings of nostalgia, which are fixated upon previous experience (and their "authentic" recreation), SSE made the past something that could always-already be experienced. Home thus becomes not something that one returns to but something that one dwells within and creates. SSE helps to make this ever-changing milieu of sensuous experience possible.

Another way to think about this is through what cognitive scientist Andy Clark (1999: 11) calls "wideware". Building on the well documented phenomena of how Alzheimer sufferers maintain a high level of functioning within a familiar space—for most their home—Clark argues that similar processes underlie how we *all* make sense of the world. In other words, we all rely upon external props and in doing this they become "part of the cognitive machinery itself" (Clark 1999a: 14). Thus, just as the world of familiar objects to some degree "speaks" to individuals with this dreadful disease, non-Alzheimer sufferers look to an array of props and aids (e.g., maps, laptops, global positioning technology, etc.) to make sense of the world.

By evoking the term "speaks" I do not mean to anthropomorphize the objective world. But it is important that we break free from the belief that subjective qualities (and ghosts) are entirely housed in the mind. Take the "softness" of the sponge. Our gut instinct would be to locate the phenomena of softness in either the sponge or the subjective experiences of consciousness. It is, rather, a quality found in the process of human-sponge interaction. Sponge softness exists in our active probing of the sponge, which varies depending upon whether we are, say, pressing it with our finger tips or rubbing it against our face. Also part of this process is how the sponge "responds" to our probing—for example, does it resist our touch or take in our fingers (Myin 2003). Every body experiences the sponge differently, just as sponges experience bodies differently (an ant, for instance, would encounter the sponge as a solid object). As Francisco Varela once noted during an interview, perception is often tied to the affordances of the body:

> Going through life as a small fly makes a cup of tea appear like an ocean of liquid; an elephant, however, will see the same amount of tea as an insignificant drop, tiny and barely noticeable. What is perceived appears inseparably connected with the actions and the way of life of an organism: cognition is, as I would claim, the *bringing forth of a world*; it is embodied action (emphasis in original) (quoted in Poerksen 2006: 37).

This gives us a way to talk about a very particular type of violence incurred as we lose biodiversity. Shelly talked about a "sense of loss" in not being able to find a particular variety of tomato that her great-grandmother grew and which, in her words, "made the best BLT sandwich ever". "I'd give almost anything to find some of those seeds", she once told me. She used to grow this variety of tomato but explained how one bad growing season caused her to lose all of her seed. "Once I realized I may never grow it again I felt like I lost a part of me, a part of my past", she eventually confessed with a sense of grief in her voice. Once we abandon the worldview that cleanly separates knower from known, mind from body, we open up a new way to talk about bodily violence (something almost universally abhorred). The pain Shelly talks about here might sound like hyperbolae; a bit of rhetorical flourish to make a point. But through the lens of wideware—a concept that allows for the ontological smearing of mind and body—we can understand this pain as more real, sensuous, and physical than might otherwise be allowed.

In *Habits of the Heart* we find reference, almost in passing, to "communities of hope". These communities "allow us to connect our aspirations for ourselves and those closest to us with the aspirations of a larger whole and see our own efforts as being in part contributions to the common good" (Bellah et al. 1985: 153). The authors argue that such "communities are not hard to find in the United States" (p. 153), pointing to ethical communities, "each with its own story and its own heroes and heroines" (p. 153), religious communities, with their scriptural stories and ritualistic practices, and a national community. It's their ability to make us feel like we're a part of something that makes these communities hope-*full*. These connectivities lie not only in a shared past. Communities of hope also tie us to a shared future.

I am not sure what Bellah and his colleagues would think of the "communities" discussed in this chapter. I have already discussed how SSE tied people to a shared past, by telling stories about seeds, educating people on traditional gardening practices, and tuning people to the tastes of generations past. Yet SSE instilled within people more than just an interest in saving the past. Respondents also expressed a hopeful narrative about the future.

Julia put it to me this way: "There's something about feeling like you're part of something bigger; a bigger movement, though I know social movement is not the right term here. But when you first see a place like this and get excited about what they're doing you know other's must be getting excited to. That makes me hopeful". These sentiments were echoed by others:

> Bill: "I've taken what I've learned here and shared it with others. […] I've gotten a few of my neighbors addicted to heirloom tomatoes and now they grow them. I'm sure others [SSE visitors] have similar experiences. That's encouraging".

> Mike: "I see people of all ages here [at SSE]. It's especially great to see the kids. […] They talk a lot about conserving the past here but really our sights are all set on the future; on making sure our grandchildren and their grandchildren have access to the same plants that we have. […] Being part of this little community gives me hope that my great grandkids might actually live in a world that's more than just acres upon acres of corn and [soy]beans".

Unlike most sites of memory, SSE is not about just the past, which, again, explains why SSE does not offer the same level of bodily reprieve as found at, say, "ground zero", the Vietnam War Memorial wall, or the Normandy Memorial. In truth, SSE is not really even about remembering, or saving, any-*thing* in particular, in the sense that it does not attempt to "fix" its objects in space and time (unlike, say, conventional gene banks). This very point is made in the organization's mission statement (quoted earlier): "Unlike Fort Knox, Heritage Farm is not surrounded by security fences and guards". Rather, their "perimeter is patrolled by Bald Eagles, red-tailed hawks, deer, raccoons and other wildlife". This blurring of boundaries between "inside" and "outside" gives visitors of SSE an active role in this process of saving. Rather than forcing individuals to return year-after-year to obtain something "authentic" SSE sees its visitors (and their gardens) as just another part of its network. As active participants in SSE, respondents expressed having reasons to be hopeful about the future. For, in the words of Mike, by "feeling like seed savers gives people a say in what that future will look like", respondents "had reasons to have a positive picture of what lies ahead".

The Performance of Forgetting

"These events are vital to the project". SSE holds seed saving seminars at least once a year. Their most popular is their Heirloom Tomato Tasting Workshop. Here, participants get to not only taste more than 40 different kinds of tomatoes but learn how to save the seeds of their favorite varieties for future planting. I am talking with Debbie, a long time member of SSE, about the value of these workshops. She continues:

> "People often mistake this place as just another seed repository. But they do a lot more than just save seeds here. [...] They should change their name to avoid this misconception. This place is also a place where people get real hands on training about things like seed saving; about how to grow these unusual varieties, how and when to pick them, how to eat them, can them, freeze them. That's what we talk about here. It's about so much more than seeds".

> "Why is this important?", I asked.

> "Because we've forgotten so much. Maybe that's why people think that this place only saves seeds. They've forgotten the rest".

Increasing attention is being placed on cultural memory for its role in conservation (Ingold 2000; Nazarea 1998). But this memory is disappearing, like the biological foundation upon which it stands. Shiva (1993: 9) refers to this collective amnesia as the "politics of disappearance", where the incommensurable is replaced with the commensurable, open-access substituted for the commodity, and diversity supplanted by monocultures. Speaking of this loss in rural American, Wendell Berry (1990: 157) writes:

> As the local community decays along with local economy, a vast amnesia settles over the countryside. As exposed and disregarded soil departs with the rain, so local knowledge and local memory move away to cities or are forgotten under the influence of homogenized sales talk, entertainment, and education. The loss of local memory and local knowledge—that is, of local culture—has been ignored, written off as one of the cheaper 'prices of progress', or made the business of folklorists.

Those interviewed believed SSE sought to resist this amnesia of the local, noting how place-based knowledge is no less real and no less essential to conservation than the seeds and the coded bits of protein beneath the seed coat. One individual put it to me this way: "If we don't know how to properly propagate them, recognizing that all seeds are different, some like direct sun, some prefer well drained soil, some hate the wind. If we forget this, so caught up saving the seeds but forgetting the other stuff, the seeds won't be much use to future generations". Not *forgetting the other stuff*: this was a common theme among respondents. This theme also appears in anthropological research studying how indigenous cultures treat their seeds, giving further understanding to what these people mean when they say "if you don't take care of the plants and talk to them and relate to them, they get lonely and go away" (Martinez 1996: 6, as quoted in DeLind 2006: 140).

In *Cultures of Habitat*, Gary Nabhan (1997: 2) describes the following notable correlation: "Where human populations had stayed in the same place for the greatest duration, fewer plants and animals have become endangered species; in parts of the country [US] where massive in-migrations and exoduses were taking place, more had become endangered". Reading this I was reminded of something Aristotle once pointed out: that although ordinary citizens lack the cobbler's expertise in how to make good shoes they still know when the shoe pinches. While Aristotle was referring to the politics of popular government, there is an important embodied lesson to be learned here: that we dwell in the world much like our feet dwell in our shoes. And just like no one knows the world of my feet better than my feet, no one knows the world of my body better than my body.

Yet, to some degree, we've lost some of that knowledge. Yes, we're very busy (from the standpoint of embodied knowledge the phrase "no longer taking the time to stop and smell the roses" is quite apt). Plus we've become socialized to believe that the truth lies not beneath our feet but up there, in a mystical ether that we can only know by stripping ourselves from the very fabric of the universe we are studying. Nietzsche (1969: 50–60), for one, was unabashedly critical of the West's quest for eternals, which had for centuries directed our attention away from life's quotidian encounters due to their perceived banality. Respondents, too, believed we've lost touch with something worth holding onto. And that corporeal act of holding onto, gripping what's around us, is important if we are to preserve the dynamic records conserved at SSE. As DeLind (2006: 134) reminds us, "knowledge that is not used, and information that is not felt, are indistinguishable from ignorance". The act of physically knowing place and doing knowledge was important to respondents. As Julia explains:

> A farm is not a test plot. Land, soil conditions, weather conditions, it's all different. That's the assumption of today's genetically engineered seed varieties. They're all the same because they assume every piece of land is the same. It's not. That's what this place is all about. It lets people conduct their own field tests. […] We need to get back in touch with the specifics of our backyards.

Talking about the premium placed upon standardized knowledge in agriculture and agri-businesses' position on saving local knowledge, Mike remarked: "I'm sure they'd prefer if we didn't know this".

"Why is that?", I inquired.

"Because by remembering we're able to remain independent. We don't have to go to them for seed and inputs. We can grow what we want; I mean what we really want and don't have to settle for their varieties. [...] Many of them taste just fine, don't get me wrong. But I like variety in my diet and in what I put in my garden every spring and I think it helps with the overall ecology of my garden. [...] I'm sure if they [agri-business] had their way they'd prefer we all forget about how to save seed".

Many respondents were like Mike, espousing the view that conventional agriculture represented a type of totalitarian regime of the mind (and body). I am using the term totalitarian regime very specifically here. This is not to outright equate, say, Monsanto with Kim Jong Il or the government of North Korea. Rather I am speaking of a totalitarianism of the mind; totalitarianism as a cognitive, lived effect. In his book *How Societies Remember*, Paul Connerton (1989: 13–14) notes that,

All totalitarianisms behave in this way: the mental enslavement of the subjects of a totalitarian regime begins when their memories are taken away. When a large power wants to deprive a small country of its national consciousness, it uses the method of organized forgetting. [...] What is horrifying about a totalitarian regime is not only the violation of human dignity but the fear that there may remain nobody who could ever again bear witness to the past.

Reading this statement called to mind Orwell's (1992: 260) deeply profound phrase from *1984*: "Who controls the past, controls the future: who controls the present controls the past". I always understood what this meant, intellectually at least. Studying SSE and talking to all those people put the phrase in practice for me. The monocultures of the mind that Vandana Shiva (1993) writes so passionately about in her critique of industrial agriculture is about the controlling of the past through the present, which, in turn, shapes future potentialities.

Corey: "My granddad would plant a lot of different stuff; having a garden with a dozen different varieties of peas was just a normal thing. [...] He'd be rolling in his grave if he saw what counts as having a 'diverse' farm these days. You're diverse if you're growing more than corn and [soy]beans".

By forgetting how biodiversity was once understood from the bodies of our ancestors, where growing (and knowing) "a dozen different varieties of peas was just a normal thing", we are far more likely to accept "what counts as having a 'diverse' farm these days". In addition, forgetting past understandings of biodiversity also narrows our collective imagination about future acceptable levels. Corey's grandfather, by Corey's own admission, went by a different measure when it came to defining biodiversity than many do today. By holding onto that knowledge and introducing more bodies to it, SSE (and other spaces like it) is creating a challenge to monoculture agriculture by refusing to make acceptable definitions that equate diversity to corn + soybeans + 1.

The Ought/Want Problem

As one staff member explained: "We try to make people actually want to save seeds". When asked how they did this they first talked of the role of explicit knowledge: "We educate. We talk about things like ecological integrity, the importance of diversity, of grower independence". Only later did they admit that "it's not something you can talk people into; we just hope people get excited about this—that's the best motivator". Sure enough; it seems to be working. In the words of the Bart: "You can't help but get excited about heritage seeds. This place makes you want to keep saving seeds if that's something you've been doing and start saving seeds if you're new to the practice".

> *Excited (noun): Being in a state of excitement; emotionally aroused; stirred* (as defined by The Free Dictionary).

Ultimately, embodied states seem to enlist desired behaviors among potential seed savers. People are not talked into wanting to plant heirloom varieties but made to feel they need to. Angela explained this feeling as follows: "I can't explain it. I come here and the next thing I know I've bought another tomato plant or a couple packets of seeds and all I can think about is what I'm going to try out next spring. […] This place makes you feel special, like we have this important role to play in saving these old plants". It is one thing to know you should be preserving seeds and the memories and practices associated with them. When I lecture about SSE in my classes I know that my students get the significance of places like SSE. Yet, for most, "getting it" rarely translates into "doing something about it". There really seems to be something in respondents' arguments that SSE "makes you want to keep saving seeds" (Bart) due to an ability to make "you feel special" (Angela) in your role as an agent of conservation. These behaviors appear to be rooted in the embodied experiences that SSE makes possible. Laura DeLind (2006: 126) once wrote how "[w]hat are needed are ways of thinking and feeling about local food that cannot be easily appropriated and/or disappeared by the reductionist rationality of the marketplace and that can balance and reframe an economic orientation with more ecological and cultural understandings of people in place". Such ways of thinking and feeling, where ecological and cultural sentimentalities are valued (even though that value may not have a dollar sign attached to it), appear to be alive and well within those whom I interviewed during my visits to SSE.

Most have heard of the so called is/ought problem. As the argument goes: One cannot get an "ought" from an "is". This thesis, which comes to us from Hume, argues that there is a class of statements of fact which is logically distinct from a class of statements of value. In other words, no set of descriptive statements can by themselves—without the addition of an evaluative premise—produce a statement of value. To believe otherwise is to commit what is known as the naturalistic fallacy. In the following chapter I dismantle this "problem" by showing it to rest upon untenable assumptions. What I'm interested in at the moment is what I will call the "ought/want problem": that merely knowing how one *ought* to act is rarely enough to elicit a particular behavior.

I see this all the time in the classroom. I lecture to my students daily about the environmental, social, and (long-term) economic costs of business-as-usual when it comes to the dominant food system. I know those students at the end of the semester know what they *ought* to do when it comes to making food purchasing and broader collective mobilization decisions. Do these clear ethical signposts lead to any type of behavioral change? As best as I can tell: no. Ethical behavior does not come from with-out—existing "out there" as naturalistic moral philosophy claims—but from *within* lived experience.

SSE afforded respondents with a desire to behave in a particular way. The term "affordance" comes from Gibson's (1986) ecological theory of perception. Affordance, for Gibson (1986: 127), is defined as "what it (the environment) offers the animal, what it provides or furnishes". To put it another way, affordances speak to how the environment *affords* the body with a variety of actions and sensations. For example, light affords the body the ability to see color, a landscape dotted with objects affords the body a sense of visual depth, and an environment of water affords the body such actions as swimming and floating (in addition to certain kinesthetic sensations unavailable to a body on dry land). But I think Gibson could have taken the concept further, by speaking of how the environment affords people reasons—wants—to act in particular ways. For example: never having walked through a forest, tasted an heirloom tomato, or heard the nighttime symphony of crickets and frogs in one's backyard, I just don't believe a brain in a vat—the archetype example of a the Cartesian "mind" guided only by disembodied reason— would ever be able to fully grasp why biodiversity ought to be protected.

I realize that people had to first want to go SSE before they could have the lived experiences made available through this space. None of the people I spoke with were there against their will. One could argue that the people I spoke with were behaviorally inclined to act in ways to protect things like biodiversity before first visiting SSE. This still does not mean SSE did not further reinforce those behaviors. The evidence in fact strongly suggests that SSE made individuals want to act in ways to further the mission of this space.

The spell of the sensuous can be quite intoxicating (see Abram 1997). Nicole explained it to me this way: "There are reasons for caring all around us. For me it's like I'm under a spell. And it's not just about the seeds. It's that these seeds, and the work of novice gardeners like me, might someday have value beyond just being beautiful to look at or good to eat. I can't think of anything more motivational than being actively involved in this conservation effort. It's no longer just intellectual for me. Being here ensures that".

Looking for Nicole to elaborate, I asked: "Could you explain what you mean by that last point? What does it mean to say that something's 'intellectual' and what does a 'non intellectual' experience look like?"

> "Non-intellectual" is not the right word. It's one thing to be told about this stuff or to read about it. But when you get to actually see this stuff, to see what we're missing, what we could lose if we don't work to save these plants, than it makes it something else, it makes it more real. It's not *just* intellectual here. It's more than that. […] That's where the power of this place lies. That's why this place, as I said earlier, has a spell on me. I've known for as long as I can remember that I should be doing this stuff but [SSE] sort of brought it all home to me.

And, again, we come across the term "more than". This time it is used to describe a powerful source of motivation. This sensual "more than" placed a spell on Nicole making her act in ways that long standing ethical arguments could not do.

The Politics of Taste

Seventeenth-century mathematician, poet, philosopher and nun Juana Inés de la Cruz observed the cognitive rush of energy whenever she was in the kitchen, leading her to write, "had Aristotle cooked, he would have written a good deal more" (as quoted in Korsmeyer 1999: 37). The power of taste is undeniable (which is no doubt why it has for so long been viewed with contempt in certain quarters). To speak personally, there are foods that have deep meaning to me. Many of these feelings originate from my childhood. When I eat them I feel like a part of me is 10 years old again. And during this experience I often remember things more clearly—an image of my younger sister sitting next to me, the feeling of that old kitchen table that I haven't seen in years, or maybe simply a nondescript smell or sound washes over me. For example, the Czechoslovakian heritage on my mother's side makes the *kolache* deeply meaningful to me. To taste this pastry filled with fruit—which in our family tends to be either prune or apricot—is to be momentarily transported to another place and time. An event made real through a dance between my tongue, taste buds, saliva, and the shedding of molecules from a familiar food. As described in Chapter 2, food has a deeply mnemonic quality to it. Lee (2000: 216) gives us a taste of this memory in a quote from an elderly Korean who, since emigrating to Japan, is unable to eat the "Korean food" she knew as a child: "My stomach and heart ache together from even just smelling Korean food, because it brings back all the hardship I have suffered in my life. Even if I wanted to forget, I cannot. My body has absorbed the past like a sponge. Forgetting is an impossibility".

"It sure brings back memories. […] They're memories but they're not. I just mean it feels real". Bruce and I are in the Visitors' Center, next to the Amish Paste Tomatoes. Talking about how this variety was a favorite in the household of his grandparents while growing up he explained, "they remind me of that time unlike anything else, pictures, stories, old things from their house". For Bruce, the memories of the past are made visceral through this six to eight ounce red fruit. Sarah noted a similar embodied realism tied to taste: "The experience is less ephemeral; when you eat you are actually feeling something you felt in the past, the experience is one-in-the-same". After talking to individuals at SSE I have a difficult time believing those who argue that "food cannot express emotion [...] [n]or can it move us in the way great art can" (Korsmeyer 1999: 109) (see Mary Douglas [1982], specifically the chapter titled "Food as an Art Form", for an alternative position). Indeed, it seems that the consumption of food moves us in ways that visual art never could; an experience that originates from within—rather than from a distance—as the categories of self and other literally spill into each other.

Taste, I would argue, is deeply political; its power to move and motivate vastly exceeds, for example, the capabilities of speech. This was documented at SSE by its ability to keep visitors interested in seed saving. In the words of one respondent: "You can lecture me about why I need to save seeds and I might do it. Get me to make a connection with a particular variety by having me eat it and like it, well, I promise I'll replant that seed every single year". The power of taste publically played out during the Flavr Savr tomato debacle. The Flavr Savr tomato was the first genetically engineered food product to reach the market. Approved for sale in May 1994 by the US Food and Drug Administration (FDA), the tomato was engineered to ripen for a longer period of time on the vine and to retain firmness longer to reduce losses during shipping. A host of variables account for the Flavr Savr tomato's failure in the market (see Pringle 2005: 68–77). But one was taste. Many consumers simply did not care for its flavor and overly firm skin (Avise 2004: 69). Had its flavor lived up to the hype, consumers and food companies (Campbell's Soups, for instance, had initially contemplated using the tomato in its soups) might have been more willing to pay the higher market price that this fruit was going for. In the end, few were willing to pay more for a tomato of inferior taste and texture.

In her essay The Site of Memory, Toni Morrison (2008: 77) notes the power of emotional memory through the metaphor of the flooding river:

> You know, they straightened out the Mississippi River in places, to make room for houses and livable acreage. Occasionally the river floods these places. "Floods" is the word they use, but in fact it is not flooding; it is remembering. Remembering where it used to be. All water has a perfect memory and is forever trying to get back to where it was. [...] It is emotional memory—what the nerves and skin remember as well as how it appeared. And a rush of imagination is our "flooding".

Industrialized agriculture has been attempting for decades to straighten out our sensibilities towards food and thus make room for its McDonaldized artifacts designed "for all seasons and all reasons" (Nazarea 2005: 11). Forgetting tastes makes things like the Flavr Savr tomato more palatable, literally, to society. The tastes nurtured at SSE represent a type of flooding; a form of remembering that does not conform to the sensibilities of industrial agriculture. The space evoked an emotional memory for respondents; a memory of and through the flesh that makes us want the type of diversity offered at SSE, rather than the monocultures of the mind and body that Global Food offers.

I am sure we can all relate to the power of emotional memory. I know I certainly can. I can recall times hearing a song that I haven't heard in 20 years and, if it was a song I listened to repeatedly back then, being able to remember its lyrics perfectly. The sound of the music and the feel of the rhythm brings it all back to me. Or ever pass someone on the sidewalk wearing the cologne or perfume of a loved one (or perhaps a long lost loved one)? Remember the rush of memories and emotions that followed? I was recently lecturing to one of my classes about Slow Food. I was explaining to them that while society places importance on saving things like biodiversity, language (though the extinction of languages has only recently entered public consciousness), and material cultural artifacts like paintings, pottery, and historical buildings, we have yet to think about food in a similar way. I could tell they still weren't buying the argument so I decided to give them a personal example. I explained to them how I felt more closely connected with my Czechoslovakian heritage than either my Irish or German roots. This connection, I told them, has nothing to do with knowledge (or more accurately lack thereof) of authentic Czech music, clothing, or language. Rather, the connection I feel to this heritage is almost entirely felt through and because of food. Even my limited grasp of the Czechoslovakian language is mostly confined to either foods (such as the delicious aforementioned *kolache*) or food related artifacts (such as *kuchenka* [cooker/stove]), which, coincidently, was most likely acquired while sitting around the table eating.

In a previous section I quoted Debbie who spoke of how SSE does "a lot more than just save seeds". Perhaps to someone unfamiliar with heritage seed banks it might appear as though they are in the same business of saving as their much larger cousins, like the earlier-mentioned National Center for Genetic Resources Preservation (NCGRP) located in Fort Collins, Colorado. It is easy to miss this "more than" that SSE saves because it doesn't reside in plastic bags or freezers. Rather, this "more than" resides in bodies, in emotional memories, embodied practices, and corporeal sensations that cannot be reduced to words or a finger point. Places like SSE seek to tune bodies so this "stuff"—since what I'm talking about cannot be faithfully represented such an imprecise word is apt—can live on for future generations.

Seeds Have Politics

Political theory often speaks of politics as involving only subjects. The vulgarities of materiality—of emotion, impulse and other corporeal activities—have no place, it has long been argued, in reasoned discourse. Along these lines, Hannah Arendt (1958: 84–5) makes a distinction between "action" and "work". The former refers to the turn-taking discourse we often ascribe to (ideal) political debate. Work, in contrast, involves those activities one engages in for survival. Arendt believed political debate ought to be insulated as much as possible from the material realities of everyday life so as minimize self-interested behavior. Jürgen Habermas (1984: 86) makes a similar distinction in his writings on communicative action, in which actors in society seek to reach common understanding by reasoned argument, consensus, and cooperation rather than through action strictly in pursuit of their own goals. Yet these views are too Cartesian, too modern (*a la* Latour [1993: 137–8]), to be of any help to us here. If the world really is smeared, as the aforementioned literature on embodied cognition suggests, then perhaps it is time to start talking about "things"—and the *relations* they affect and are an effect of—as having politics too.

In a paper titled "Where are the Missing Masses? Sociology of a Few Mundane Artifacts", Latour (1992: 153–5) writes about, among other things, doors. Latour sets out to explore the potent effect of doors—and other technological artifacts—upon social order. By speaking of the "missing masses" (a word play upon the black matter—the missing mass—in physics), he argues that conventional analyses of order miss an important component: the power of things. Technological artifacts, like doors, shape human behavior. These material artifacts, by bringing forth new relations, produce an *effect* that is often indistinguishable from normative, moral, and legal control. Some examples: a door shapes the speed and direction we can walk through walls; a speed-gun fastened to a stoplight enforces certain traffic codes similar to a human policeperson; and a seat-belt that automatically slides over a person after ignition implements seat belt laws. In technology, then, lies the ability to delegate actor-like status to these material artifacts, which, in turn, means we delegate power to them; making then, literally, powerful.

Latour continues by discussing what is known in France as the "groom"—the hydraulic door-closing device attached to many office doors. The groom shapes how we move through the door—pulling the door closed too violently forces people to behave differently than if the door closed slowly. The eccentricities of a particular groom become what Latour (1992: 158) calls a "local cultural condition", whereby those who use the door regularly acquire the skill to move through it unscathed while those who do not get caught in—or potentially hurt by—its unconventional actions. Door-closers thus discriminate against the very young and the elderly (who might move through the door too slowly), as well as against anyone carrying heavy, bulky, or long items (such as delivery people, furniture movers, etc.). Elsewhere, Latour (1999: 187–8) writes about the governing effects of the "sleeping policeman" (the speed bump). The speed bump governs the speed at which people drive. Speed bumps thus produce effects as if a real policeperson (state control) or one's friends and neighbors (normative social control) were standing along the street. But rather than a fine (state control) or shame/public humiliation (normative social control), speed bumps operate physically—by inflicting damage to a car's suspension.

Bruno Latour and others of the same theoretical ilk (like Donna Haraway, Michel Collon, and Isabelle Stengers, to name but a few) have spent their careers attempting to up-end our conventional view of the world and its seemingly clear divisions. The limitations of this worldview are no better illustrated than in the tiresome debate over whether guns or people kill people. Given the assumptions of Western political thought, with its focus on states rather than relations, we only have these two choices. But clearly a gun cannot shoot a bullet without the involvement of a human and a human cannot shoot a bullet without the involvement of a gun. The framing of the debate misses a third interpretation, involving a "citizen-weapon" or "weapon-citizen" (Latour 1999: 179). Thinking in terms of relations allows us to see that a citizen-with-a-gun-in-hand is no longer the same as an unarmed-citizen, just as a loaded-weapon-in-hand is no longer the same object as an unloaded-gun-in-a-safe. In short, a smeared worldview recasts action as a relational effect. In Latour's (1999: 182) words, "action is simply not a property of humans *but of an association of actants*".

Stepping outside a perspective that focuses on the identification of fixed subjects and dead objects—and adopting instead a relational view—makes it less necessary think in terms of exclusions (Jullien 2002). When the "really real" is a fluid and ever-changing process dichotomistic and exclusionary logic suddenly appears naïve. As Wood (2005: 166) explains: "Where the yes/no border logic is dominant, it often reflects an underdeveloped capacity for thinking, that is, for negotiating complexity, or the recognition that there are forces that would disempower those who think in such a way". Exclusionary logic drives everyone to the extreme. Following this logic, it *is* either Y or Z (the question of whether guns or people kill is an example of this underdeveloped capacity for thinking). There is no third way; no middle ground to choose from.

Being cannot be determined outside of its network. Embodiments therefore not only shape our knowledge of the world, they also populate that world with "things". A seed-locked-in-a-cryonic-chamber is not the same as a seed-possessed-by-people-who-know-how-to-grow-it. The latter seed, because of its network-determined-being, is of greater consequence than the former seed, which many never see the light of day (and if it does no one may be in possession of the requisite knowledge to put it to work). Steven made this point well when he explained:

> Gene banks don't do an adequate job in my opinion because they remove too much. They couldn't care less about protecting the knowledge and memories that go along with knowing how to grow those seeds and knowing what conditions those seeds grow best in. [...] Gene banks might talk about saving seeds, just like they do here [SSE], but they're not the same.

People change, too, with changes to their networks. That's the other point that Latour et al. are attempting to make, namely, that our own state of being is predicated upon network configurations. And as those associations change so too changes who and what we are (Haraway 1991: 150). Let's return to Shelly, who talked earlier about the "sense of loss" experienced after losing a heritage tomato variety given to her by her great-grandmother. Later in that conversation she remarked: "I might have physically been the same person [after losing the variety] but I just didn't feel the same. […] I just felt, somehow, like something was fundamentally different. I mean, I know what changed, I didn't have the seed anymore". In other words, Shelly-with-the-seeds-of-a-heritage-tomato is not the same as Shelly-without-the-seeds-of-a-heritage-tomato. Thinking of being in these mutable terms casts spaces like SSE in a new light for it reframes what it is that these spaces do. When being is viewed as fixed and immutable, then the "bank" analogy would adequately describe SSE, where *things* are put in and later taken out. Yet when being is understood *relationally*, SSE is recast as doing a form of conservation that far surpasses that done by more conventional gene banks, which reduce being to, in the words of an earlier-quoted individual who works at NCGRP, "protein, information, and water".

SSE could therefore be viewed as a site of resistance. More than just event-*full*, it is also power-*full*. For Foucault (1980: 236), power and resistance are not held or acquired but reside within the connections among things. Power is therefore not something possessed but something *produced* (see e.g., Foucault 1980: 97). Power is practice; it is a relational effect. You therefore can't escape power relations, according to Foucault, because all relations are literally power-*full* (which is to say relations are by definition *productive*). At best, one can only hope to change relations. This is precisely what SSE attempts to do. By changing peoples' lived experiences with food SSE is seeking to change their understandings of what food is and how it ought to be produced and consumed. When viewed through a Foucaultian lens SSE could be understood as a site of resistance to the dominant discourses, practices, and knowledges associated with Global Food. As Lori remarked: "[SSE] is a game changer. Looking at it from the outside you might only see people and seeds. From the inside, however, it's all about building connections and those relationships have a real effect".

Chapter 5
The Sensibilities of Chicken Coops

In *Totemism*, Levi-Strauss (1963: 69) criticizes the functionalist explanation given to the human-animal relationship offered by British Anthropologists Malinowski and Radcliffe-Brown. Levi-Strauss argues that "natural species are chosen not because they are 'good to eat' but because they are 'good to think'" (Levi-Strauss 1963: 89). More than "creatures that are feared, admired, or envied [...], their perceptible reality permits the embodiment of ideas and relations conceived by speculative thought on the basis of empirical observation" (Levi-Strauss 1963: 104). In other words, non-human animals—which I will shorten to "animal"—are good to think with. To be clear, I agree that animals think. And yes, it is possible that animals find humans good to think with too. But I cannot, nor do I wish to, speak for animals, agreeing with Haraway (2003: 1–10) that speaking for animals falsely gives the appearance that they are doing the speaking. Our unavoidable humanity, as the basis from which we know the world and ourselves, requires that when I speak of the epistemic affect of animals I do so from the standpoint of a human body (Johnston 2008: 643).

In *If You Tame Me*, Irvine (2004: 7) writes: "our relations with animals, on the cultural level, have as much to say about human beings as about animals themselves". While sympathetic to projects that attempt to think about animals sociologically, I worry that statements like this are too generalizing—assuming Culture with a capital "C"—viewing human-animal relations as homogenous across society, which in turn produce shared understandings about animals and ourselves. Perhaps there was a time when this was the case, such as when ours was an agrarian society and most everyone lived with animals. But today's embodiments—at least as far as our lived experiences with the animal world are concerned—vary tremendously. For example, Yarwood and Evans (2007: 107) illustrate how breeds of livestock become intertwined with geographic regions and help in the maintenance and recreating of local identities. In his study of Turkish pigeon handlers in Berlin, Jerolmack (2007: 874) explains how their "animal practices" help produce understandings of ethnicity, culture and territory. Elsewhere, Bell (1994: 199) discusses how the fox is a source of group and self identity—and part of what Bell terms their "natural conscious"—for certain members of the village Childerley.

But these relations also shape how we think about animals. Though long dismissed as an effect of "primitive" cosmologies (see e.g., Durkheim 2001: 167), social scientists are beginning to take seriously the knowledge of those who continue to know animals in an everyday, lived sort of way. Making the case for these embodiments, Ingold (1988: 16) explains:

As far as I know, it is quite unknown for people with this kind of experience to endorse the psychologists' view of their animals as mindless, unthinking machines. Ought not this fact give the psychologists pause? Is it not the business of a science to produce views which are supported, not contradicted, by the most testing forms of experience which involve their subject-matter, never mind if such testing takes place outside laboratories?

For Ingold and others (see e.g., Haraway 2003; Hinchliffe et al. 2005; Johnston 2008), the potential to know with and about animals may offer an access point into the type of de-centered cosmology that places "concern for" at an even level with "knowing about", the latter being of utmost concern in Western thought (Latour 2004b: 225–30). Ingold (1994: 19) talks of how this could represent a "starting point" in the "history of human concern with animals, in so far as this notion conveys a caring, attentive regards, a 'being with'". He continues by "suggesting that those who are 'with' animals in their day-to-day lives, most notably hunters and herdsmen, can offer us some of the best possible indications of how we might proceed" (p. 19). But why must we go "back"—at least that's the imagery I have when Ingold writes about "hunters and herdsmen"—to find instances of being with animals in their daily lives? Granted, for most, being with animals today has radically different meanings and practices associated with it than compared to just a couple generations ago. These changing patterns of human-animal spatial relations deserve some attention.

Being with Animals: A Brief Historical Overview

A significant body of literature already exists detailing how human-animal relationships have changed over the centuries, with a number of analyses going back to the first cases of animal domestication (see e.g., Anderson 2004; Bulliet 2005). My interest lies in more recent trends, when humans and animals became separated into distinct social worlds as the latter were relegated to fewer and fewer places where they were considered "in place". Following the seminal work of Mary Douglas (1966), societal definitions of "pollution" are culturally variable and reflect societal views of defilement or disorder. This led to her now famous definition of pollution as "matter out of place" (Douglas 1966: 35). Public understandings of pollution thus reflect socially defined moral boundaries—that is, of what is (and is not) natural and thus right. These moral boundaries are defined spatially, in terms of phenomena being "in place" (e.g., the good/the right) or "out of place" (e.g., defilement/the bad/the unnatural). Douglas' argument has since been extended to speak to, for example, the politics of social exclusions (e.g., Cresswell 1996; Sibley 1995), odor perception (e.g., Bubandt 1998), and public understandings of air pollution (e.g., Bickerstaff and Walker 2003). But this conceptual frame is also useful for understanding our changing relationship to animals.

Chris Philo (1995: 664–76) provides a useful starting point with his analysis of how animals (save for a few notable exceptions) were slowly excluded from city spaces in the late eighteenth/early nineteenth centuries. This shift in public understandings of animals within city spaces had links to public health discourses as well as to fears of moral degeneracy. In terms of the former, health reformers were looking to cleanse the filthy urban environment, with the emanations of animals (e.g., odor, manure, etc.) being at the top of their list of prime offenders. Regarding the latter, there was a fear that humans could become morally debased if they were exposed with frequency to the unregulated urges of the uncivil beast. Philo (1995: 666) frames these discourses by setting them against two long-term intertwined shifts:

> one involving the long-term process whereby all sorts of phenomena have become categorized in certain ways and allotted to certain spatio-temporal containers, thus raising the difficulties of what to do with "matter out of place"; the other involving the equally long-term splitting apart of the urban and the rural as distinctive entities conceptually associated with particular human activities and attributes (the industrial and civilized city, the agricultural and barbarian countryside).

Thus began the process of animals' extrication from urban environments; a respatialization that affected our understanding of them. Previous to this, animals and humans regularly lived in close proximity to one another. Not only did humans and animals share urban environments but animals readily transgressed boundaries between the "in here" of the home and the "out there" found beyond one's doorstep. Centuries ago it was not uncommon to find, for example, chickens, goats, and pigs scurrying about one's home, particularly among the peasant class. It was not only peasants who allowed such transgressions to occur. Kings too were known to favor having animals in their place of residence. The recent discovery of three lion skulls in the Tower of London points to the location of the Royal Menagerie backing back to the twelfth century. In addition to lions, the menagerie is known to have held an elephant (although only briefly), ostriches, bears, leopards, and tigers (Owen 2005). But a social reorganization of animals was underway. By the early nineteenth century, not only was it becoming unacceptable to allow animals into one's home (save for "pets"), it was also becoming increasingly objectionable for animals to be located near large populations of people—hence, their expulsion to the countryside. In both cases, within the "in here" of one's home and the "in here" of the city, animals had become "out of place"; this, even though only decades earlier they were considered "in place" in both spaces.

The countryside, in many respects, thus represents the last bastion of space— save for notable exceptions, like zoos, stockyards, dog parks, and the like—where animals could still be considered "in place". In recent decades, however, animals have undergone yet another significant respatialization. These more recent trends threaten to make animals "out of place" even in the countryside. The cause of this most recent movement can be traced, at least in part, to a reorganization of agriculture and a parallel respatialization of non-farmers into rural areas.

The Changing Moral Geography of Animals in Agriculture

The industrialization of agriculture can be traced back to the nineteenth century and involves, in part, the substitution of fossil fuels and capital intensive production technologies (mechanical, chemical, and most recently biological) for labor. This process has led to the consolidation, concentration, and specialization of farms as farmers seek to spread production costs over as large an operation (and as many units produced) as possible (Buttel et al. 1990). The hog industry (along with poultry) has been at the forefront of consolidation trends. While the number of hogs in the US has changed little over the last century, the number of farms raising hogs has decreased precipitously. For example, the number of farms producing hogs in the US has declined from almost 700,000 in 1980 to approximately 78,000 in 2002. At the same time, the percentage of hogs being raised by the largest sized farms (greater than 5,000 head inventory) rose from 20 percent of total US production in 1992 to approximately 50 percent in 2000. Indeed, in the six-year period between 1993 and 1999 there was a 250 percent increase in hog operations containing 5,000 or more animals (USDA 2004).

As livestock become concentrated in fewer and fewer areas, many other areas are becoming sanitized due to the disappearance of these animals. Such a trend is the effect of two concurrent processes. First, there are fewer farms today than a generation ago, which places greater distance between neighbors. And, secondly, among those farms that do remain, fewer are raising livestock today than was the case a few decades ago. Our ancestors, particularly those that resided in the countryside, lived in very close proximity to animals (Anderson 2004). This is becoming less the case today. For example, poultry has gone from being raised on approximately 78 percent of all farms in the US to just 4.6 percent during the latter-half of the twentieth century. For hogs, that figure, during the same time period, went from 56 percent to less than 4 percent of all farms. Dairy cows likewise went from being raised on 68 percent of all farms in mid-century to only 4.3 percent of all farms 50 years later (USDA 2004). And, lest we forget, not only is livestock raised on a much smaller percentage of farms today than was the case a couple of generations ago, but there are significantly few farms *in total* (making these figures all the more remarkable).

These trends—namely, the changing structure of agriculture and the populating of the countryside with non-farmers—have played an important role in shaping understandings of livestock and, I would argue, animals more generally. Moreover, these trends have created a conception of rural life remarkably devoid of the animals found on a livestock farm; a conception of rural life where livestock (e.g., hogs, cattle, sheep, etc.) are increasingly becoming "out of place". Indeed, given their close coupling to agriculture, one could argue that such animals can only be deemed "in place" in those spaces where agriculture too is "in place". As mentioned earlier, for centuries this described the countryside. Yet, as agriculture becomes increasingly "out of place" in the countryside—as a result of its disappearance from this space—so too will follow those animals bred specifically for such ends.

The Case: Introductions and Analysis

The chicken once held a position in the household comparable to the television today—almost everyone had at least one. Chickens were a necessary part of any farmstead. A book from 1852 titled *The American Poulterer's Companion* explains how chickens pick up waste that "might escape the pigs", which means "much of the refuse of the kitchen can again appear on the table in a new and better form" (as quoted in Sheasley 2008: 80).[1] They could be found in cities, towns, and, of course, the countryside. Chickens were a familiar part of almost any landscape up until the twentieth century. Explained in the 1855 book titled *The History of the Hen Fever*, this "extraordinary mania" toward owning and raising chickens touched "Kings and queens and nobility, senators and governors, mayors and councilmen, ministers, doctors and lawyers, merchants and tradesmen, the aristocrat and the humble, farmers and mechanics, gentlemen and commoners, old men and young men, women and children, rich and poor, white, black and gray" (Burnham 1855: 5, 21).

Yet as in the case of cows and pigs, the twentieth century witnessed restrictions in terms of where chickens were considered "in place". A change is underway, however. Unlike other food producing animals, chickens have witnessed a reversal of fortune of late, finding themselves "in place" in more places today than was the case just a decade ago. Well over a 100 US cities have code in place making it possible for an individual to raise at least two chickens on their property. The question "Why is this occurring?" is beyond the scope of this chapter and book, though it would make for a fascinating sociological study. My interest, following earlier chapters, centers on questions of knowledge: specifically, what are the epistemic effects of these chicken-human relations as they relate to understandings of self, animals, and food?

To investigate this, individuals were interviewed who raise chickens within (or near) cities along the Front Range of Colorado; specifically, the cities of Boulder, Fort Collins, Longmont and Loveland. Interviews occurred between September 2008 and March 2009. A total of 21 individuals were interviewed. Each interview lasted approximately one and a half hours. All interviews were tape-recorded

---

1   Early on, chickens were kept for their eggs, not flesh. The popularity of poultry at the dinner table is a relatively recent phenomenon, made possible by the industrialization of chicken production. For example, our ancestors ate, on average, nine pounds of chicken in 1928, significantly less than the 60 pounds of poultry we reportedly each consumed in 2008 (Sheasley 2008: 86).

and later transcribed. None of the interviewees raised chickens or eggs for sale nor did any self-describe themselves as "farmers". While a few (n = 5) were raised on a farm, no one had raised chickens their whole life. The experience of owning chickens, for all, was relatively new, spanning anywhere from 12 years to two months. Individuals known to myself were first approached and asked to participate in this research. Those individuals were then asked to indentify others who own chickens within or near the city limits who might be interested in participating in this research. Interviews were semi-structured. I also watched, observing participants in their casual interactions with chickens. This helped ground my findings, for it opened the interviews up to questions that emerged out of these situated encounters. These observations also allowed me to compare formal responses with *in situ* behaviors, such as whether respondents' often-touted care and respect for their birds actually meshed with their lived, observable practices. Interviews continued until similar themes began to emerge, which is to say, until "theoretical saturation" was reached. As in previous chapters, the names of participants have been changed to protect their identity.

A conceptual aside: from the standpoint of embodied knowledge, backyard chicken coops are, at least it seems to me, a sensual resource that cannot be compared to, say, the occasional visit to either a petting zoo or livestock farm. The repetitive animal-human interactions that are part of raising chickens in one's backyard create a phenomenologically rich lived experience. These practices create certain understandings—of, for example, "chickens", the categories of "animal" and "human", and "food" more generally—that cannot be obtained from books alone. This is why I chose such an admittedly unusual case study. (I remember the strange look I received after telling a colleague that I was studying human-chicken relationships.) Also, because of its recent moral geographical turn—where chickens are finding themselves "in place" in more spaces than a couple years ago—I was curious to find out if backyard chicken coops offered society anything more than just fresh eggs and compost material. As I learned, this relationship was the source of so much more than food and fertilizer.

Minding Chickens

Marc Bekoff speaks of "minding" animals in two ways. First, he uses the term to refer to caring for animals. The second meaning refers to attributing a mind to them, which involves "wondering what and how they are feeling and why" (Bekoff 2002: 11). While the concept is meant to blur the lines between what it means to be human and nonhuman—an ontological smearing with clear ethical consequences—I worry it overstates the cognitive capacity of the human animal. In his Forward to Leslie Irvine's book *If You Tame Me*, Marc Bekoff tells of how in his study of coyotes he becomes a coyote; "likewise", Bekoff states, "with my research on dogs and birds" (Irvine 2004: vii). I wonder if he would ever think it possible for coyotes (or dogs or birds) to become human. If not, these attempts at creating what Deleuze (1995: 240) calls a flat ontology mistakenly have just the opposite effect: they privilege what it means to be human. My point is not that we cannot think like chickens, coyotes, dogs, or birds. We simply cannot know what it means to think like these nonhuman animals and therefore have no basis upon which to make such statements.

Only being able to think like a human does not, however, mean we cannot still mind animals. We just have to change our understanding of the concept a bit. To do this let's return to Andy Clark's understanding of the mind. In his book *Being There*, Clark (1999b: 53) tells of how the mind is a leaky organ that does not remain confined within the envelope of the physical features of flesh and bone (Ingold [2009: 153] later offers the corrective that the skull is leaky and the mind is what leaks). This same point was made decades earlier by Gregory Bateson (1973: 429) when he proclaimed that "the mental world—the mind—the world of information processing—is not limited by the skin". To speak of "minding animals", then, means developing practices whereby these creatures become part of an unfolding system of relations, action, and ultimately perception.

George Herbert Mead is perhaps most responsible for giving (human) language the pivotal role it has had in social psychology. For Mead (1934: 268), language, or to be more precise the vocal articulation of words (what Mead called "significant symbols"), is what separates humans from non-humans. Animals communicate, according to Mead, but that communication is preprogrammed and reflex-like in its nature—a dog bark and a cat hiss are stimuli that produce the behavior of a "back off" response in other dogs and cats (Irvine 2004: 120–21). Unlike in Mead's time, there is now a mountain of evidence debunking the belief that the animal kingdom lacks complex communication systems. But all of this talk about, well, *talk* seems to miss the mark when it comes to human-animal communication. This brings us back to the concept of minding animals.

"I know when they're sick. They tell me; not with words, but they do. You just got to be open to hearing them, to the signs". Jed has kept chickens for four years. He also grew up with chickens, on a farm in Nebraska. He clearly cares for his birds. You can tell by how he talks about and to them. In his words, "they're part of the family".

"What do you mean you have to be open to hearing them? How do they tell you they're sick?", I asked.

"After a while you get to know each bird, I mean really know them. You get used to certain behaviors, certain actions. When those behaviors change that's a red flag. The signs are often pretty subtle but they're there. That's what I mean you got to be open to them. Things like swollen joints, diarrhea, and changes in their movements, in how they hold their wings. [...] Because we don't have many birds you get to know what you have pretty quickly, seeing them as often as I do. So when something happens you notice it right away".

For Jed, his chickens "tell" him when they are sick, though not through verbal cues. This highlights the importance of other, non-verbal, forms of communication. Though perhaps unusual for someone with more modern sensibilities, people have for millennia been "listening" to the behavior of chickens for signs. Augury is the art of divination through the observation of birds; a practice said to be invented by the Ancient Romans, who looked to birds to predict the outcome of military and political battles (Stanley 1920: 495). A century ago scientific justifications began to replace mystical explanations for aviary observation, pointing to how "birds are susceptible to atmospheric changes (of an electrical and barometric nature) too slight to be observed by man's [and women's] unaided senses" (Stanley 1920: 495). When talking of augury today, scholars use the term to reflect upon embodiments that are becoming increasingly rare in Western countries. It represents a reminder of "a type of knowledge-making (about the present and the future) that is in danger of disappearing in the twenty-first century: the knowledge gained by intimate relations with animals" (Squier 2006: 70). I think we can push the point still further, in terms of arguing that birds "spoke" to some of the respondents.

Research into how Alzheimer's patients "communicate" to their caregivers reveals interactions similar to those described by some of the respondents. Holstein and Gubrium (2000: 137), describing earlier research on the Alzheimer-caregiver relationship, refer to this communication as follows:

> Caregivers may persist in "articulating" the victim's mind long after he or she has lost any capacity for self-expression. Contrary to Mind's view that the individual expresses him or herself to others, the Alzheimer's experience finds others literally speaking and "doing" the mind of the victim as a way of preserving it. Mind thus becomes a social entity, something interactionally assigned and sustained, both by and for whomever assumes it to exist.

Caregivers, in other words, get a feel for the mind of the other, showing, as Bateson argued, how the mind is not limited by the skin. The same was true for many of the respondents, as they described communicating with their birds without the aid of verbal communication. In another interview, Betty talked about "sensing" how her birds are doing: "I can't really put my figure on it. And it's not one thing either, but how they move, respond to me, and interact with the other birds". Later in the interview she added how her "chickens communicate [to her] with sounds and motions".

Just like I could never interpret an Alzheimer victim's mind after a single visit, respondents were not able to initially "hear" their chickens. In the words of Bart, "it wasn't until after a couple weeks that I saw intention in their actions; before that it looked like plain old chaos to me". In short, minding takes time—an effect of repeated face-to-beak interactions. And with this comes empathy. This makes sense given how such communication makes whatever we are minding appear a little more human than before. Shapiro gives a thoughtful account of this process—what he calls kinesthetic empathy—as he talks about understanding his dog, Sabaka. According to Shapiro (1997: 277), kinesthetic empathy "is possible because we both have a living, mobile, intending body". This notion of empathy involves the non- or pre-linguistic—that of smells, sounds, tactile sensations, sights, and the like (recall Betty mentioning how "chickens communicate with sounds and motions")—which draws humans and animals closer together. For Shapiro, this non- or pre-linguistic communication takes the form of meaningful looks and behaviors. Respondents, as the following quote reveals, discussed a similar process in acquiring an empathetic understanding of their chickens: "You get to a point where these animals mean something to you. After awhile, you can't help but care for them". Thus, while this togetherness does not eradicate the human/animal distinction it does appear to reduce the likelihood of seeing animals in strictly instrumentalist terms.

Ethical Living From the Ground Up

Betty, whom we heard from earlier, explains how being with chickens has shaped her ethical orientation to these animals:

> I don't condone eating eggs from major egg companies but I understand why people are apathetic about where their eggs come from. The well-being of chickens is down on the totem pole for most people. 'Chickens aren't the same as cows or hogs—they're just stupid birds!' I get why people think that way. Most have never had any real interaction with a chicken before. If they had they wouldn't think that way.

Philosopher Hub Zwart (1997) argues that the Cartesian view of animals, which reduces them to amoral machines, has helped legitimize their objectification and sustained the view that they are resources for human use. In opposition to the Cartesian view, philosophers and social theorists talk about a dwelling perspective—or what Ingold (1994: 19) refers to as "being with"—in an attempt to up-end the grip offered by Global Food (see, e.g., Johnston 2008; Thrift 1998). Elsewhere I have written on what I call the "epistemic distance" of sustainable agriculture (Carolan 2006c). This speaks to how many of the benefits of sustainable agriculture cannot be immediately experienced—for instance, increases in soil fertility take time to occur. Conventional agriculture, conversely, has witnessed such success in part because its benefits are immediate in comparison to its costs—for instance, its ability to externalize costs makes its food cheap compared to the alternatives. Related to this is the fact that food production as a whole has become epistemologically distant from the average consumer. This distance benefits industrial food because it creates gaps and distortions in terms of the information available to consumers, which increases the likelihood that it will exist in a readiness-to-hand state.

  While most people have driven by a field and seen vegetables in an unprocessed state, far fewer have experience being with living, breathing livestock. It is hard to believe that the spatialization trends discussed earlier, whereby livestock are increasingly only "in place" where humans have no place (many of today's meat processing centers strictly regulate who's allowed through their doors), has been of no consequence to how we feel about animals that feed us. Discussing the expanding logics of capitalism, rationalization, and techno-scientific knowledge, sociologist Zygmunt Bauman (1991: 193) speaks of the "vanishing point of moral visibility". I believe Bauman is speaking about how our present economic system makes it possible to act with moral indifference; a system largely invisible to us erodes some of our moral grounding toward processes that make it possible.

I am using the term "grounding" here quite literally. There is a long tradition in the West of separating ontological claims from moral claims. As the argument goes: one can't get an "ought" from an "is". This thesis, which comes to us from Hume, argues that there is a class of statements of fact which is logically distinct from a class of statements of value. In other words, no set of descriptive statements can by themselves—without the addition of an evaluative premise—produce a statement of value. To believe otherwise is to commit what is known as the naturalistic fallacy. Yet this position, which after centuries continues to ruffle philosophical feathers, is premised upon a particular metaphysics: one that stubbornly clings to a world rejected by the phenomena of embodied knowledge. The logic of "cold", abstract ethical reasoning is premised upon the very divisions that a lived view of knowledge nullifies. Note Immanuel Kant's position on the subject: "No moral principle is based [...] on any feeling whatsoever. [...] For feeling, no matter by what it is aroused, always belongs to the order of nature" (quoted in Okin 1989: 231). Yet, as already established, reason divorced from feeling is simply unattainable. As Ingold (2000: 25) points out, "an intelligence that was completely detached from the conditions of life in the world could not think the thoughts it does".

What, then, does this tell us about ethical reasoning? If one accepts that we do not think as minds-in-vats but as bodies-in-the-world then one must be likewise willing to accept that we think ethically in a similar manner. What happens when the logic of "cold" ethical reasoning, moving out from first principles, produces results that are offensive? If so-called objective reasoning produces an ethical principle that we find reprehensible will we suddenly accept that moral posture simply on the basis of its sound logic? Of course not. We accept those moral principles that *feel* right and discard those that go against what, in our "gut", we know to be wrong (Ingold 2000: 25).

Kennan Ferguson (2004: 374) notes that while few ethicists would make the argument that one ought to save dogs over refugees, all dog owners do precisely that whenever they take their pet to the veterinary clinic. They are making a choice that *prima facie* appears, through the lens of disembodied reason, unethical. Yet I hear no calls for the closing of veterinary clinics until the world's refugees are given adequate water, food, clothing and shelter. Taking another swipe at the logician's mind (without a body), Ferguson (2004: 385) explains: "It may well be that, logically, those who eat meat should indeed have no compunctions about eating dogs, even their own dogs. Of course, such an argument will prove attractive only to those whose affinity for logic exceeds their affinity for dogs". Varela (1999: 1–10) differentiates between what he calls ethical know-how and ethical know-that. Drawing upon phenomenologist Merleau-Ponty and pragmatist John Dewey, Varela is essentially talking about how that *feeling* between what is right and wrong develops. Rather than being encoded in what is known as theoretical/linguistic knowledge, these feelings arise through practical, embodied skills (what Bourdieu [1995: 93] terms "bodily hexis") that cannot be reduced to drawings or language. This ethical know-how literally grounds our feelings about how we *ought* to act in the world.

It is interesting that moral arguments were originally employed to increase the distance between humans and animals, such as the belief that animals represented a moral danger to city populations due to their noisy presence and unrestrained animality (most notably their openly sexual behaviors) (Philo 1995: 660–65). By altering what Donna Haraway (1991: 237) calls our "situated moral understanding" of these creatures, this moral regulation also changed our ethical relation to these animals (Smith 2002: 50). Losing this previously held grip on these animals has made it harder to empathetically relate to them.

Alison was introduced to backyard chickens through her partner, Jack. Before meeting Jack she never raised chickens nor did she, by her own admittance, think much about where her eggs came from or the chickens who laid those eggs. In her own words:

> [Jack] was doing this when before we met. I would have never dared try this on my own. I was never much interested in growing my own food. Other than growing pots of herbs I just preferred going to the store. [...] Never thought much about eggs [before meeting Jack]. Never really cared, if I'm going to be honest with you. [...] I think about that stuff now [talking about the well-being of chickens]. Once you get to know these animals, being around them every day like I have, you don't see them the same. At least I don't.

When I talked to Alison she had been with her chickens for about eight months. During that time, Alison acquired a moral understanding of chickens that was shaped significantly by being around them every day, to the point that she no longer saw them the same as she did before meeting Jack. This point was also conveyed in her actions. Alison certainly appeared to care for her birds. The voice she used while talking to them was much like one might use with a child. The way she looked at them, often while broadcasting a smile. Alison genuinely cared for those animals. As I wrote in my field notes: "Shows genuine empathy towards her birds by her actions and in how she speaks to them".

Alison was far from alone in expressing this notion of a grounded moral understanding as it applies to the feeling respondents had toward their chickens.

Steven: "Chickens are totally mischaracterized in our culture. They're not the same as rocks, contrary to what some of friends tell me. [...] I wasn't always fond of these birds but I am now".

Claire: "They [chickens] always interested me, obviously or I wouldn't have gotten into this. But I don't think I can say I always cared about them, not enough at least to, when I have to buy eggs at the store, to be willing to pay a little more for the free-range type".

Sally: "Having chickens in my backyard has brought me closer to these birds in ways beyond mere proximity".

These quotes highlight the limitations of disembodied, "cold" ethical reasoning. Not only do we think through our bodies, we also think about what we ought to do through our bodies. One of the problems with Global Food is that it denies us certain embodied ethical resources. As Claire explained, "If you've never been around chickens I can see why you won't think twice about their well being. I mean, how close to chickens can someone feel when their closest relationship to one is the Chicken McNugget?" If one's embodied knowledge of chickens is based significantly on encounters with Chicken McNuggets why should anyone expect them to feel ethically connected to this fried mass of bleached meat? We shouldn't.

More than just a representational understanding of chickens, respondents possessed a *relational* understanding of these animals. They showed how embodied practices do more than transfer meaning. These practices *are* the meaning.

As a whole, sociologists tend not to find much of sociological significance in the body. As a Durkheimian social fact, the body has been viewed largely as something that is inscribed upon—by, for example, culture, language, discourse, and power. Rarely has it been viewed as that which makes social facts possible. This assumption has been famously referred to by Ingold (1995: 66) as the "building perspective": "Here, then, is the essence of the building perspective: that worlds are made before they are lived in". Or to put it another way, the building perspective, to borrow a phrase from Merleau-Ponty (1992: 24), denies the simple fact that the body is the homeland of our thoughts.

Discussions of how we come to carry forms of dwelling, involving skills, sensibilities and dispositions, sound to me a lot like Pierre Bourdieu's (1995: 93) concept of "bodily hexis". Bourdieu (1995: 14–18) makes the important observation that, beginning in our earliest years of life, we become inculcated to certain social practices, tastes, attitudes, and understandings. This unconscious feel for the game of social life appears to us as natural and objective but it is in fact a product of specific embodied relationships. In Bourdieu's (1995: 95) words, the bodily hexis becomes "a permanent disposition [he's overstating its rigidity here], a durable manner of standing, speaking, and thereby feeling and thinking". Preceding Bourdieu, we find Norbert Elias (2000) making similar observations, most notably in his magnum opus *The Civilizing Process*. Elias (2000: 141) details at length the manner in which feelings, dispositions, and tastes—specifically in the context of changing definitions of civility and good manners—are ground within material practices that over time are internalized, becoming "second nature". Elias coins the term "sociogenetic" to give us language to discuss how socio-historical practices (and their corresponding feelings and beliefs) are passed down from one generation to the next, to some degree like genes.

Our feelings about animals, which too have a seemingly self-evident quality to them, are grounded in socio-historical practices. Is it thus any surprise when surveys report how animals, and livestock in particular (see, for instance, Swabe et al. 2005), exist for segments of the population in a readiness-to-hand state—as things/resources that satisfy needs rather than entities in possession of needs themselves (Kruse 1999; Pifer et al. 1994)? Global food has transformed social space, particularly in terms of how we relate—cognitively, epistemologically, and morally—to animals. For those born into this space, these social practices, and the dispositions, feelings, and tastes therein fostered, appear objective and self-evident. As the food system is presently arranged, it is easy to not think about those animals that enrich our dinner tables. As Stan explained: "Never having been a 'farm boy' I really didn't give much thought to livestock. I just cared about food, cheap food basically, because I never really knew that end of food [livestock production]. It was only after getting into chickens that I began to think about the animals at the other end of the system". Betty made a similar observation: "Most of my friends and colleagues live a life devoid of any interaction with animals, especially animals like cows, pigs and chickens. They don't interact with them, they don't have any reasons to, so they don't think about them, even when eating meat or their morning eggs. That stuff's not an animal product. It's just food for them".

During my interview with Steve we got on the subject of when he began to feel differently about his birds. Steve, in his own words, "only recently became a chicken guy". This was a somewhat unusual topic of conversation. Not wanting to force respondents to pin down exactly when they noticed changes in their dispositions, feelings or understandings toward food, I intentionally avoided asking questions on the subject. Steve led me into this topic by noting specifically when he knew the birds had changed him.

> I had a sick bird; she was off her feed. I couldn't get her to take anything. I remember going out there a lot to check on her. Kept trying different things to get her to eat—cat food, creamed corn—but she wasn't interested. One of my buddies was over. I guess I was noticeably bothered by the experience. He said something to me, like "what's the big deal, it's just a bird". Something like that. And for a second I thought, "what are you talking about". For a second I didn't process that this behavior was a little weird. I mean, from the perspective of someone from the outside. It seemed perfectly natural to me.

It seemed perfectly natural to me: the empirical equivalent to saying "this is my bodily hexis". This statement is all the more significant because it came from someone who, as mentioned earlier, claimed he "wasn't always fond of these birds". And what changed to bring about this effect? As best I can tell: his relationship with chickens.

Nigel Thrift (2004: 60, emphasis in original) once wrote, "*affect is understood as a form of thinking*, often indirect and nonreflective, it is true, but thinking all the same". Stated another way, the relationship *itself* is a living, fluid, non essential social fact; it is something that's sociologically real (and sociologically consequential) but which is all too often "cancelled out by such methods as questionnaires and other such instruments" (Thrift 2004: 60). In sum: *how* we know the world is as important as *what* we know, because the former ultimately informs the latter.

Ambivalence and Eating Friends

Unlike a lot of livestock, backyard chickens don't usually end up on the dinner table. Most city ordinances allowing chickens within the city limits only allow the animals for egg production, not for meat. But some respondents, living outside city limits, did occasionally kill their chickens for consumption. Dwelling with these animals did not turn respondents away from eating meat or eggs. I do not want the aforementioned empathy and care conveyed by respondents toward their birds to be conflated with, say, the empathy or care one would give an infant. It is a gross oversimplification to believe that an increased phenomenological grip on animals will translate into a vegan or vegetarian lifestyle. I am aware of no relationship between owning dogs and rejecting the taste of cooked animal flesh. I've never met a rancher who didn't eat meat. Similarly, I know a number of vegans and animal rights activists who have very little first-hand knowledge of the animals whose rights they're working to champion. In sum, I know of no research that has found a correlation, either positive or negative, between "being with" animals and the consumption of their meat. That being said, respondents did talk about how their dwelling with chickens created conscious tensions in how they thought about the meat products they consumed.

Others have noted the tensions of knowing, caring for, and eating animals. Humphrey (1995: 119), for example, explains how "we can hold multiple, even seemingly contradictory attitudes to the very same animal". This point is illustrated by the following saying from the Nuer (a pastoral people in southern Sudan and western Ethiopia): "The eye and the heart are sad, but the teeth and the stomach are glad" (recorded by Evans-Pritchard 1940: 27; quoted in Tuan 1984: 92). John Berger (1980: 5) described a similar sentiment when he exclaimed: "A peasant becomes fond of his pig and is glad to salt away its pork. What is significant, and is so difficult for the urban stranger to understand, is that the two statements in that sentence are connected by an *and* and not by a *that*".

Not surprisingly—given the obvious fact that they ate eggs—none of the respondents self-identified as vegan. What was surprising, however, was the ambivalence each respondent expressed towards meat eating. The word "ambivalent" derives from the Latin prefix *ambi*, which means "both", and "valence", which is derived from the Latin term *valentia*, which means "strength". Contrary to its popular usage, ambivalence is thus not the same as indifference. When someone is ambivalent, they possess feelings, attitudes, and beliefs that are in tension with each other. The term was coined in 1911 by Eugen Beuler to describe the contradictory feelings that accompany schizophrenia. As it is used here, attitudinal ambivalence is defined as having simultaneously both positive and negative evaluations towards an object, issue or behavior (Berndsen and van der Pligt 2004: 71).

Ambivalent attitudes have been found to be less stable and result in more reflexive behaviors (Berndsen and van der Pligt 2004: 71–2). Looking into attitudes of meat consumption, Sparks and colleagues (2001) found that ambivalence was strongly correlated to moderate attitudes about food consumption, likely due to conflicting motives about what they eat. Berndsen and van de Pligt (2004) reached a similar conclusion, finding that higher levels of ambivalence towards a particular diet—from eating meat regularly to veganism—weakened the link between attitudes and the dietary practices they actually followed (see also Povey et al. 2001).

Respondents expressed their ambivalence towards meat consumption in a variety of ways. Some were quite philosophical about it. In a statement reminiscent of Tuan's (1984: 9) proclamation that "Eating [...] is an expression of love [...,] [w]hat we love we wish to incorporate", one respondent explained:

> It's tough not to sound hypocritical, I know. I say I really care about these birds but then I kill them. In what world does that constitute as care, I know. I think of it this way. These birds and I have formed a bond, a pretty close bond. When I eat them that bond continues, even grows stronger because they're feeding my family. By eating them we're taking them into our family.

Another respondent, perhaps as a strategy to reconcile this cognitive dissonance, made light of these contradictory attitudes through the use of humor: "They're like my pets. We've got names for each of them. [...] Maybe that's why Jack [their dog] keeps running away. He's probably worried we're going to eat him too!" And for still other respondents, nothing was done to reconcile these contradictory beliefs: "It's something I frequently think about. I can't explain away this tension. How can I? It is what it is. I'm aware of it, though. And I think it's made me more aware of what I'm eating in terms of how it is raised".

It's something I frequently think about [...]. It's made me more aware of what I'm eating in terms of how it was raised. Statements like this point to practical suggestions for those looking to change food consumption patterns. Unlike the concept of indifference, ambivalence implies a degree of reflexivity. To be ambivalent about something means you are *aware* of conflicting attitudinal positions. The global food (and livestock) system, it could be argued, seeks to create indifference. Backyard chicken coops, it appears, have the opposite effect. Berndsen and van de Pligt (2004: 77) note the power of ambivalence for changing meat consumption patterns when they explain:

> Our findings could also have important practical implications for people who are involved in the promotion of change in food consumption. In line with Povey et al. (2001), we believe that ambivalence could play an important role in promoting changes in meat consumption. Decreasing people's ambivalence, for instance, by creating more favourable viewpoints about the affective and moral aspects associated with meat consumption, should increase meat consumption. Conversely, if one intends to reduce meat consumption, one should increase people's ambivalence, by stressing the negativity of affective and moral aspects of meat consumption.

In light of what I learned from respondents, I would like to refine one point made by Berndsen and van de Pligt. Ambivalence can shape more than the *quantity* of meat individuals consume. It appears to also play a role in making individuals more aware and interested in the *quality of life* that livestock have. For all but two respondents, raising backyard chickens had *no* effect on how much meat and eggs were consumed on a yearly basis. Of those two one reported eating fewer eggs "to keep the cholesterol in check" while the other stopped eating chicken after becoming "a little too fond of them". Being with chickens seems to have caused respondents to give more thought to the ways their food (particularly where animals are involved) was produced. Mark put it to me this way: "I can't say being around chickens has changed my diet. I still love my animal products. [...] It's fair to say, though, that I no longer just eat this stuff without any thought to how the animals were raised. I think about food a lot more now". While for Ralph "the act of daily coming into contact with the birds makes it next to impossible to eat a chicken product without at least briefly thinking about what's on the other end of the fork".

Along with the technical and economic efficiencies created through industrialized animal production are also attitudinal efficiencies. Global Food has been tremendously successful in (and its success is also because of) its ability to make us forget and to not ask questions about the origins of our food. And for good reason. As the literature on ambivalence and food consumption indicates, once we begin to reflect upon what we eat we become more apt to change our dietary patterns (see, for example, Jonas et al. 1997; Povey et al. 2001). The experiential effects of chicken coops should not be dismissed. While further study is needed, the phenomenological grip offered by such "being with" experiences as those described by respondents points to the yet-to-be-recognized value of backyard chicken coops for those looking to challenge Global Food.

Managing Risk

Chapter 3 discussed how respondents viewed their participation in CSA as involving a substitution of uncertainties, where unacceptable uncertainties (e.g., chemical inputs) were exchanged for acceptable ones (e.g., composition of weekly food share). Involvement in CSA brought a greater level of discursive consciousness; an awareness that elicited intrapersonal reflection and discussion with others about chemicals in agriculture. Similarly, respondents for this chapter appeared highly sensitive to the risks that accompany industrial scale egg, poultry, and meat production (which furthers contextualizes their aforementioned ambivalence toward the consumption of these commodities).

> Steve: "I know what I got here. I know my eggs are fresh. I don't have to worry about not knowing about if the eggs are old or covered in E. Coli".

> Nick: "I don't trust what I get at the store. You have no idea what you're getting. I can't say it's not safe but I can tell you that what I'm raising behind the house is".

> Barb: "You have no idea what those birds are fed or what the processing facilities look like [referring to industrial scale egg production]. I don't have those worries here [referring to homegrown eggs]. [...] Not only do I think my eggs taste better but I also think they're better, safer, for my family to eat".

Respondents did not, however, claim that raising backyard chickens was free of uncertainty. Rather, as in the case of CSA consumers, it was a matter of substituting unacceptable uncertainties for acceptable ones. Mark put it to me as follows: "I can't say I like not knowing just how many eggs I'm going to have from one week to the next but at least I know, of the eggs that I do have, they're fresh and that they're not going to make anyone sick". Lisa made a similar comment: "I gladly accept the various uncertainties that go along with raising chickens. [...] Sometimes your birds die unexpectedly or a raccoon or fox will get at them. [...] Thanks to industrial egg production I can be certain that [a local grocer] will have eggs whenever I want them and that they're all of a certain size, shape, and color. But there are still a ton of unknowns that go with eating store bought eggs. I'd rather know the conditions under which my eggs were raised and gathered". Barb echoed these concerns: "I may not know what will be out there on any given day [referring to number of eggs laid] but at least I can be confident when I'm eating eggs that they were raised and handled in ways that are best for not only the chickens but for my family too".

Sociologist Ulrich Beck (1999: 12; see also 1992) has written at length about how non-knowledge is an inevitable product of the expansion of capital and techno-scientific rationality. According to Beck's thesis—what is famously known as the "risk society thesis"—this unawareness allows for the global flow of information and technologies. Largely unforeseen in all of this, however, are the associated risks that accompany this expansion. According to Beck, risks today are an ever-present and largely unavoidable by-product of modernity with its associated technological and organizational forms. Take the case of Global Food. While production efficiencies and scales of economy make food today remarkably inexpensive, they also presuppose an organizational form that is inherently risky. As slaughter facilities continue to push to increase their speed and efficiency mistakes will inevitably happen. Increasing production line speeds increases the risk of cutting open internal organs while the animals are being eviscerated thus exposing flesh to fecal matter. As the meat from different animals becomes mixed in processing facilities, so increase the chances of cross contamination as the bacteria of one animal suddenly finds itself spread across thousands. Accidents, scares, and risks are therefore a "normal" (Perrow 1999) part of today's complex food system.

Another term is helpful here: agnotology. Coined by the historian of science Robert Proctor (1996), agnotology refers to manufactured ignorance. A prime example of agnotology is the tobacco industry's decades-long public relations assault that involved producing doubt about the cancer risks of tobacco use (Michaels and Monforton 2005; Proctor 2006). But agnotology need not only refer to ignorance induced by lies, deceptions, and distortions. More generally, agnotology serves to alienate us from other ways of knowing and alternative epistemologies that may challenge conventional patterns of social organization. In this sense, agnotology is very much at work in our knowing—or non-knowing—of Global Livestock. And, like CSA, the embodied knowledges acquired while *doing* backyard chicken coops appear to induce understandings that bring that ignorance into the light. This point was articulated clearly by Claire. In her own words:

> As someone familiar with animals you learn quickly that the old machine analogy is dangerously incomplete. You can't treat animals like a machine and expect to produce a safe product in the end. You just can't. Animals get sick, they carry pathogens, they shit. Machines don't do this. And when you slaughter an animal, or pick its eggs or milk it you risk exposing those things to a host of other things that could get people sick or maybe even worse. Handling hens on a daily basis educates you. They're animals, not machines.

Claire is speaking of a trend to reduce livestock to machines; a linguistic reduction that goes back well over a century. In the book *Chicken World, Profitable Poultry Production*, published back in 1910, M.G. Kains tells the reader to think of the hen as another piece of machinery. When calculating production costs one must "not only [think of] the cost of the egg as a market commodity, but the cost of making the machine, the hen, which is to manufacture the egg" (Kains 1910: 18, as quoted in Squier 2006: 76). A non-chicken example where the machine trope is used to describe livestock can be found in the book *Old MacDonald's Factory Farm*, which explains that "the breeding sow should be thought of as, and treated as, a valuable piece of machinery whose function is to pump out baby pigs like a sausage machine" (Coats 1989: 32).

As an example of agnotology, it is not that the animal-as-machine trope sets out to outright lie or deceive. But, in practice, that is what it does. That's Claire's point. If animals really were machines then many of the risks associated with large-scale industrial animal protein production would not exist because, to paraphrase Claire, machines don't get sick, carry pathogens, or shit. Working with chickens on a daily basis showed Claire, and others, the mistakes of this analogy. And it helps explain why respondents viewed their eggs—and, as implied by Claire, most locally produced forms of animal protein—as a safer alternative to what is available at the local grocery store.

Making Problematic "Cheap" Eggs

Writing about the history of the production of "freshness" as it relates to food, Susanne Freidberg (2009: 65) explains that consumers had to first be convinced "not only that fresh beef could come from far away, but also that their main relationship to meat—and indeed, to all once-living foods—was *as* consumers" (emphasis in original). She then likens this shift in how we relate to once-living foods—having shifted away from possessing at least some relationship to the animals and those who raise them and prepare their flesh—as an example of commodity fetishism. This transformation has changed our understanding of food from that of a process, which requires things like labor and natural resources, to that of an "it". I made it a point in every interview to bring up the costs of raising one's own eggs. Eggs in Northern Colorado from a chain-grocery store cost on average a dollar a dozen (at my last trip I noticed them on sale for 88 cents a dozen!). This easily makes buying eggs from a Safeway or Super Wal-Mart cheaper than if one were to attempt to raise egg laying chickens in their backyard. At minimum, respondents reported having a couple hundred dollars invested in their backyard chicken operation, while a couple respondents claimed to have spent well over a thousand dollars. As my economist friends tell me, it simply doesn't pay to raise your own eggs. So why do it?

All respondents reported that the cost of industrial egg production is not fully reflected in the price listed in the grocery store. As explained by Steve: "cheap eggs aren't as cheap as you think". I realize I have, both figuratively and literally, a chicken and egg question here. Clearly respondents must have felt this way before investing the time and effort in raising chickens. So I recognize it is untenable to attribute the lived experiences of raising chickens to views that question the "cheapness" of industrial eggs. Each respondent admitted to recognizing the costs of cheap food well before deciding to raise chickens. Nevertheless, each respondent also reported how raising chickens strengthened that belief, which translated, for a number of individuals at least, into measurable behavioral changes.

Stan, who is earlier quoted as saying he cared only about acquiring "cheap food" in his younger days, decided to raise chickens for their fresh eggs and to give his children "something to do on the weekends". He knew that he would, in his own words, "never be able to raise eggs as cheaply as those sold at the store". But cost was not the primary motivating factor for Stan. As he explained to me, "letting my kids learn a little about animal behavior and about what goes into raising the foods we eat—you can't really put a price on that". His view on the matter seemed to change once he actually started looking after chickens. In his own words:

> "As I got more interested in my own chickens I started thinking more generally about those big industrial egg laying facilities; not just about the chickens but also about the people that work in those conditions, they [the conditions] really have to be terrible. After a while I started to seriously revaluate why those eggs at the store are so cheap".

> "What do you mean by '*seriously* revaluate'? What makes this time 'serious'?", I asked.

> "I mean, even though I only thought about price for the longest time [when making food purchasing decisions], I wasn't stupid. I knew there was something suspicious about just how cheap food has become. How could you possibly sell eggs for below a dollar a dozen and not be screwing someone? But, I'm almost embarrassed to say it, I didn't care. I still bought those cheap eggs. What changed is that my backyard acted like a peephole into this system that I was ignoring. It's easy to forget about how food is raised when you're not doing it. I'm not saying everyone that raises chickens sees their coop in the same way. But that's the effect it had on me".

In sum: Stan reports feeling differently about food since he started raising chickens. That's interesting; though a change in an individual's attitudes and beliefs is somewhat superficial if not followed by measureable behavioral changes. So: were these changes in understandings "shallow" or "deep"—that is, were they of any sociological consequence (Carolan 2007: 1265)? I'll let Stan answer the question: "I never have paid as much for eggs in the past as I do now. Fortunately it doesn't happen often but when we need to buy eggs we either get them from another local grower or we make sure that they're certified free-range and organic".

Betty was also quick to point out why we can no longer afford to buy cheap industrial eggs and how this view has only been strengthened since having poultry on her property. In her own words:

> My eggs probably cost me about a dollar each [after factoring in the initial investment]. It's not something everyone can afford, I realize that. But let's put that fact to the side for the moment. Do you honestly think we can afford to continue to destroy the environment while we also commit the type of animal welfare and human rights sins that we're committing in the name of "efficiency"?

Every day I get up and am reminded of all the work that my chickens and I do to make eggs. How could I, how could anyone, when being reminded of this daily, think those big egg laying facilities are not cutting corners? And who's paying for these cut corners? We are, in terms of things like water pollution and poor working conditions creating jobs only illegal immigrants would possibly take.

What Betty seems to be talking about, like Stan, is the need to embed eggs in a moral economy. The moral economy is a term used by economists to describe the original state of all economies before their shift into a market economy. Adam Smith (1976) was arguably the first to provide a sustained sketch of this shift. Yet Smith's ideas have become hijacked. Contrary to the strictly utilitarian view of economics usually ascribed to him, where raw selfishness and radical individualism is believed to create the most optimal ends, the vision he had of the market remained rooted in care and sympathy. For Smith, friendship, mutuality, and a moral order constituted the basis of markets (Adi 2005). My economist friends, who tell me repeatedly that it doesn't pay to raise one's own chickens for their eggs, are thinking about scales of economy. Betty and Stan, conversely, are thinking about scales of a moral economy. For them, after factoring in all the "corners" that Global Food cuts—corners made more visible after having chickens in their backyard—it doesn't pay to eat industrial eggs.

Chapter 6
Cultivating Communities

Much of the book thus far has been directed at how, particularly from the consumption side, we've become tuned to Global Food and how things like CSAs, heritage seed banks, and backyard chicken coops make its experiences, understandings, and tastes slightly less palatable. I hope the analysis up to this point has given the reader a deeper understanding and thus appreciation of these artifacts. In addition to their economic impacts (Bagdonis et al. 2009; Brown and Miller 2008; Hancharick and Kiernan 2008), phenomena like CSAs, heritage seed banks, and backyard chickens coops have been shown to improve regional, community, and household food security (King 2008; McCullum et al. 2005) and positively affect biodiversity (McNeely and Siebert 2002; Thrupp 2000). In creating an alternative grip on food, these spaces challenge Global Food in ways that have yet to be fully appreciated by agro-food scholars.

The focus thus far has been largely on consumers, in terms of how we taste, perceive, and understand the food we eat. Perhaps a book could stand alone by merely describing how our grip on what we eat has changed over the last century and how things like CSA, heritage seed banks, and backyard chicken coops offer lived experiences that challenge Global Food. To end the book here, however, would cause us to miss still other ways that these spaces exist as sources of meaning and knowledge. It is not my intention to focus solely on consumption.

Beyond merely wanting to display greater analytic balance, in terms of being able to say something about both "sides" of production and consumption, lies a more practical concern about how we can bring people into contact with what they eat in ways like those described in the previous three chapters. Even if we can agree that these spaces are experientially important, the discussion remains in the category of "wishful thinking" without people *capable* of making these experiences possible. If we want to talk about making consumers, quite literally, feel differently about food, we had better be sure there are individuals capable of producing food that fits with these alternative tunings. This chapter looks at how the spaces described in earlier chapters serve as repositories for embodied food *producing* knowledge.

The Agricultural Ladder is Being Kicked Away

The number of farms in the US hit a peak of close to seven million in the mid-1930s, followed by a decline that only recently stalled. The rate of decline was most rapid in the 1950s and 1960s, and continued to fall until the 1990s, when the number began to level off. In fact, the last decade has witnessed a rise in farm numbers. What the United States Department of Agriculture (USDA) calls "lifestyle" farms, smaller farms with more diverse commodity profiles and where the operator's primary source of income tends to come from off the farm (because farm sales total less than $2,500), and larger farms, those whose sales exceed $500,000, are on the rise in the US. This hollowing out of the middle—of middle-sized farms—is commonly referred to as the bimodal (or dual) structure of agriculture (Rogers 1988: 203). The US had approximately 2.205 million farms in 2007, up from 2.16 million in 2002 and 2.11 million in 1992 (USDA 2007). Where are these "new" farms coming from? And, just as importantly, where are those embodied repertoires of knowledge (aka farmers) learning this craft?

There is a suite of research in rural sociology that describes what is known as the "agricultural ladder" (see e.g., Bates and Rudel 2004; Becket 1969; Lyson 1979; Spillman 1930). This literature details, "rung" by "rung", the journey individuals take to become full-fledged farmers, from unpaid family work, to wage labor, to tenant farming, to a mortgaged farm, to finally full ownership of a farm (for a critique of this framework see Kloppenburg and Geisler 1985). My own personal experience of having friends who were raised on farms and eventually took over the family operation has shown me that the total amount of "rungs" might have varied. Yet for all one was universal: the rung of unpaid family work. All of my childhood friends who now commodity farm—who are, in other words, engaged in large scale industrial agriculture—grew up on a farm. Indeed, of all the conventional farmers I know there is not one who wasn't raised on a farm.

This chapter is not about the agricultural ladder. Like the other chapters, it is about embodied knowledge. The agricultural ladder, and more importantly the rung of unpaid family work, hints at the embodied knowledge required to be a successful conventional farmer. Those who think commodity farming is easy ought to try their hand at it. Contrary to pejorative statements that, for example, liken conventional agriculture to "farming out of a can" (to emphasize its reliance upon chemical inputs), industrial scale producers rely daily upon previously accumulated tacit knowledge. My childhood friends who now farm began to accumulate this embodied knowledge at a very young age. And while a couple went on to college to get degrees in fields like dairy science, animal science, and agronomy before returning home to help run the family operation the knowledge amassed while on the farm *doing* (conventional) agriculture never left their sides.

Yet having a 5,000 acre corn-soybean operation or 500 head dairy operation is not equal, embodied knowledge wise, to having, for instance, a CSA or an operation that sells produce at a local farmers' market. I would venture to say that my friend with 1,200 head of Holsteins, 600 of whom are milked three times daily, could no more successfully run another friend's 5 acre vegetable farm (who sells his produce at local farmers' markets) than the latter friend could manage the former's large dairy operation. This is not to say that one is more difficult to operate than the other. Each has its own knowledge demands, both explicit and tacit. And knowledge of how to *do* one does not translate all that well into knowledge of how to *do* the other.

It's pretty clear to me where the larger farms of the aforementioned bimodal structure are coming from. For the most part, they are operators who have managed to remain on the "agricultural treadmill" (Cochrane 1993: 427). Having climbed the agricultural ladder, their strategy for remaining in agriculture is grounded in the "get big or get out" logic made famous back in the 1970s by the then-Secretary of Agriculture, Earl Butz. So where are these lifestyle famers coming from? Some no doubt climbed the agricultural ladder, choosing, after reaching the "top", to abide by logic other than "get big or get out". A number of these individuals, however, were not raised on the farm, taking instead alternative routes into farming.

It is not my contention that these smaller agricultural forms rely upon embodied knowledge more than those lying at the other end of the scale continuum. What I am interested in, rather, is the processes by which this knowledge—especially the more "sticky" (Hipple 1994) variety—is transferred across bodies and over social channels among those producing for local food systems. It is imperative that we better understand how food production knowledge is not only disseminated but also how bodies are made receptive to this knowledge: that is, how are they made to *want* to become farmers in the first place? Like Wendell Berry (2007: 4–5), I have worries that, given the West's collective detachment from the land and food production, we risk reaching a point where we'll have insufficient embodied and explicit knowledge to raise food under any logic other than "get big or get out".

"How can this be?", you might ask. "How could we ever forget how to farm!?" As I've indicated repeatedly, not all knowledge is the same. Some forms are more mobile than others, which is another way to say knowledge differs by degrees of "stickiness". The more sticky the knowledge, the more practice is required to preserve it. Quite literally, when it comes to sticky knowledge, if you don't use it you'll lose it.

In an article published in the prestigious journal the *American Journal of Sociology*, Donald MacKenzie and Graham Spinardi (1995) argue that nuclear weapons are becoming *uninvented* as a result of global nuclear disarmament trends and nuclear test ban treaties. It is their contention that "if design ceases, and if there is no new generation of designers to whom that tacit knowledge can be passed, than in an important (though qualified) sense nuclear weapons will have been uninvented" (MacKenzie and Spinardi 1995: 44). Could we one day find ourselves in a situation where we've essentially uninvented small-scale, local food production, only knowing how to "farm big", while all the rest, following Earl Butz's advice, have gotten out? This is an important empirical question that has yet to be systematically examined by agro-food scholars.

The Case: Introductions and Analysis

While interviewing individuals for the three aforementioned studies I came across a handful of people who raised food for local sale. There were, of course, the four CSA producers interviewed for the research described in Chapter 3. While conducting research for Chapter 3 I also interviewed a customer who raised and sold sweetcorn, strawberries, and asparagus to sell at farmers' markets. There was also an individual who hoped to start her own CSA in the near future, using her position as a customer to learn more about farming. While conducting research at SSE I spoke to three individuals who grew produce for a local farmers' market. Then there was the individual who raised sweetcorn and lettuce for a local grocery store and who occasionally sold their sweetcorn out of the back of their pickup truck on the side of the road. Of the backyard chicken coop owners I spoke with, three sold their eggs at local farmers markets and to local restaurants (located outside city limits they were not bound to city codes that make this activity illegal).

An "n" of 13, I realize, might raise the eyebrows of some of the more methodologically-minded readers. To address this, and to expose these 13 to the same methodological instrument, I attempted to re-interview them during the summer of 2009. Eleven agreed to be re-interviewed over the phone. (One backyard chicken producer from Chapter 5 and one farmers' market participant from Chapter 4 turned down my request for a follow-up interview.) The shortest follow-up interview lasted 35 minutes, the longest 75 minutes. The empirical foundation upon which this chapter is grounded thus amounts to, on average, over two hours of interview material for each participant, or approximately 26 total hours of recorded data. The goal of these follow-up phone interviews was to develop a further understanding of how CSA, heritage seed banks, and backyard chicken coops served to educate bodies about how to *do* small-scale local-market-oriented agriculture. Respondents' names in the quotes below have been changed to protect their identity.

Replacing the Ladder with an Initial Shared Lived Experience

> I got into this because of my dad. He had a big garden and sold a lot of his stuff to neighbors and friends. Even plants. I remember, he'd take his car to work in the spring and right there, after work, with the trunk up, he'd sell tomato and pepper plants and flowers. […] He was the one that starting taking me to farmers' markets. Met a lot of great people during that time; have a lot of fond memories of that. […] Can't say that I knew then that I wanted to do something similar but those experiences certainly got me interested in becoming a farmer.

This quote comes from Lyle. Lyle raises vegetables for a local farmers' market. Like the other 12 individuals who are the focus of this chapter, Lyle was not raised on a farm. I found this curious. My experience growing up had been that those who produced food for sale were raised in a family that also produced food for sale. What happened to the agricultural ladder, I wondered. What caused these individuals to enter into agriculture, given that, unlike my aforementioned childhood friends, they were not born into a life of farming?

For Lyle, this grip on small-scale, local market oriented food production was nurtured through earlier experiences at farmers' markets. This knowledge, however, was not something that could have been effectively communicated to him in the classroom. The earlier experiences of participating in farmers' markets were what led Lyle to later grow food for this venue. As Lyle explains, it was also his experience that friends of his only began to find the lifestyle interesting after they began to experience it firsthand too. In the words of Lyle: "I've tried talking people into participating [in farmers' markets]. It's not that it's a hard sell. The hard sell is telling someone who has never been to a farmers' market why they should be involved in one. I really think you've got to go to one before you really get it".

This was a theme that came up repeatedly during the interviews: that earlier experiences with local food production were linked to people wanting to later participate in similar food production regimes as growers. The experience, in other words, was both a messenger and a motivator. That was what Lyle was getting at when he talked about farmers' market participation being only a hard sell to those who have never experienced this space firsthand. Yet after having this experience, "you really get it". "It", in this case, represents the non-communicable aspects of farmers' markets that make people want to participate in these spaces as both growers and consumers. Others ascribed equal importance to this "it" when describing where their interest in becoming farmers came from.

Jeff : "I started stopping at fruit and vegetable stands back when I was in college on my way back from home on the weekends. Once I graduated from college I moved to a medium sized city that had a couple good farmers' market and a few CSAs. [...] I don't know, somewhere along the way I guess I thought that it would be pretty cool to do something like that too. [...] The longer I was involved [as a farmers' market and CSA customer] the more attractive the thought of being a farmer became".

Susan: "Grew up in a small town so I've been around agriculture my whole life. Helped friends pick rock growing up and with detasseling (corn). [...] Never thought for a second that I'd be a farmer when I grew up. That came later. [...] Our family liked to garden so I always knew I had a green thumb. It wasn't until after college when I started to hang with people involved in the local food scene that I started to give real thought to being a grower myself. I really thank them for turning me on to this. I'd probably still be programming computers otherwise".

Mark: "I had known about CSA for years but it really wasn't until I joined one that I began to think that this might be something I could do".

As the above quotes indicate, there were no ladders leading respondents into small-scale, local production regimes. Respondents, rather, found their way into these alternative modes of farming by way of a number of different routes. Interesting, however, was the fact that all routes involved *previous lived experienced with local food production*. As Mark explained, it wasn't enough just knowing about CSA to make him want start one up himself. It was ultimately the experiences acquired through his membership with a CSA that made him begin to ponder it as a potential career option.

This is not to suggest that respondents located their final decision to start growing food for local markets to any one specific encounter with small-scale local producers. There is a long history in the social science literature to locate novel meanings, and thus alternative behavioral paths, within "the" event. Dating back a half century to Anselm Strauss' work studying career path trajectories, leading to what have become known as "turning point experiences" (see e.g., Becker and Strauss 1956; Strauss 1959), social scientists continue to look to the novel to explain seemingly dramatic shifts in behavior, belief and identity. This literature includes talk of epiphanies (Denzin 1989), fateful moments (Giddens 1991), and transformational experiences (Wainwright 1995). Such terms refer to "those key moments that bequeath enduring impressions on a person's life" (Wainwright and Turner 2004: 319), thus, "having had this experience, the person is never quite the same" (Denzin 1989: 15). As Giddens (1991: 112) explains, these "fateful moments are highly consequential for a person's destiny". In addition to an absence of ladders, I came across nothing to suggest that respondents' decisions to start growing food for sale were the result of any one particular event.

The adoption and diffusion literature gives us a way to think about why the aforementioned experiences might have been important for respondents in their decision to participate in local food regimes. The theory of adoption and diffusion, for those unfamiliar with it, highlights the importance of social networks in diffusing an innovation among members of a social system (Rogers 1995: 10). Traditionally, "innovations" are defined in this literature as either technological artifacts, such as hybrid corn (the artifact that gave rise to this theory almost three-quarters of a century ago) (Ryan and Gross 1943), or novel ideas, like new policies (Mintrom 1997). Yet an innovation could equally apply to anything new or novel, like a novel career or hobby.

"Mass media channels", so the diffusion of innovation literature argues, "are more effective in creating knowledge of innovations, whereas interpersonal channels are more effective in changing attitudes toward a new idea, and thus in influencing the decision to adopt or reject a new idea" (Rogers 1995: 36). Without knowing it, the adoption and diffusion literature is making a distinction between tacit and explicit knowledge in statements like this. Information conveyed through media channels is largely explicit in its form and content; it is knowledge, by definition, that has been reduced to codified form. As previously mentioned, tacit knowledge, because it is sticky, does not travel well, particularly when the method of exchange involves words and images. Tacit knowledge travels best through practice; through doing, literally, the knowledge. I do not doubt that interpersonal communication helps reduce some of the friction of this exchange. When encountering something new—whether it is a technology, idea, or perhaps even a perspective career or hobby—we like to talk about it. We ask questions. But there is more. While long understood as a theory of communication (Rogers 1995: 10), there is an implicit embodied component to the adoption and diffusion framework. According to this literature, trialability and observability are traits that increase the likelihood of adoption (Rogers 1995: 16). Both involve "being with" the novelty under investigation. In the language of this book, if people can physically engage with an innovation, and thus acquire embodied knowledge about it, they are more likely to alter their attitudes towards it. Communicating about and "being with" local food production were described by respondents as important factors in making them want to "adopt" the small-scale, local market oriented farming lifestyle.

What follows is an exchange with a respondent I'll call Jim. It came out of a conversation we were having about how his previous involvement in CSA shaped his decision to raise food for local markets.

"The bug [to become a farmer] definitely wasn't planted until I joined [the CSA]. Once I started to give it serious thought, you know, started to think that I might really want to do this, I started asking a lot questions [at their CSA]. I started going there more; helped out with picking and delivery, trying to learn whatever I could. [...] Talking to [the operators] you could just tell they loved what they did and the more time I spent in the gardens the more I realized I'd love it too".

"What do you mean that the more time you spent in the gardens the more you realized you'd love operating a CSA too?", I asked.

"Well, it's just being outdoors, working with your hands, meeting new people, being part of something that you're passionate about. It's a bunch of things, I guess. I just felt good, really good, whenever I was out there [at the CSA]".

Note the role that communication and physical engagement played in Jim's decision to start growing food for local consumption. Talking with other growers certainly played a role in this, which arguably could be replicated in spaces outside of CSA and farmers' markets. But communication alone does not a farmer make, at least in Jim's case. His physical presence in the space proved to be a powerful force, making Jim want to become a farmer too. And unlike the information acquired from talking to other producers, such embodied experiences can only be acquired within these spaces of local food production.

Following Sticky Knowledge of Food Production

Farming requires a host of knowledges, skills, and expertise (Carolan 2006a). Sustainable farming practices in particular have been described as requiring additional—more locally grounded—knowledge forms, beyond that demanded by the "productivist modes of thinking" (Wilson 2001: 77) embedded within conventional agriculture (see e.g., Hassanein 1999; Kloppenburg 1991; Winter 1997). This argument hinges on the point that sustainability requires knowledge of what's sustainable—culturally, economically, and ecologically speaking—for any particular space. From the perspective of farm management, this means knowing things like local soil types (Ingram 2008), pest ecologies (Röling and van de Fliert 1994), and pasture properties (Hassanein and Kloppenburg 1995). In her study of farmers in England, for example, Ingram (2008) wrote about how soils were often discussed in very corporeal terms, in terms of being "light and easy" or "heavy" (p. 221). Later she describes how "[t]his feel for soil" (p. 222) emerged for many farmers over time, through "experience, long-term observation and record keeping" (p. 222).

We must be careful, however, and not take this to mean that sustainable (or local) agriculture requires more tacit, practical, and embodied knowledge than conventional forms of food production. If one accepts the premise that tacit and explicit knowledge are two interconnected sides of the same epistemological coin, then we'll want to think twice before claiming that some farming arrangements are more dependent upon embodied knowledge than others. Based upon discussions from previous chapters, we ought to agree that farming, in all its forms, is premised upon a variety of different knowledges, from the most sticky to the most mobile. This begs the question: how does this knowledge, especially that which is sticky, travel, particularly with regard to new bodies that are only beginning to be tuned to the needs, understanding, and skills of local food production?

Knowing the economic realities of large scale industrial farming, it is pretty clear to me why my childhood friends who are now engaged in fulltime commodity agriculture climbed similar rungs to get to where they are today. The agricultural ladder is, at least in part, an artifact of agriculture's capital intensity (Harris 1950). In the case of most of my commodity farming friends, either they inherited the farm outright or, to provide their parents with some financial security (and make sure other siblings received a fair share of the inheritance), they entered into an attractive (e.g., reduced interest) lease-to-own arrangement. And as they climbed this ladder, brought on in part by financial necessity, they learned about farming.

As someone that studies agriculture and food systems for a living I have grown used to the litmus test used by producers—particularly those involved in commodity agriculture—to assess one's legitimacy and knowledgeability on the subject of agriculture. The question I'm most often asked upon first meeting a producer is: "So, were you raised on a farm?" Initially that question bothered me. Not having been raised on a farm, I worried this fact would somehow diminish my chances to access the target population around which much of my early research centered. Mentioning that I grew up in a small town in rural Iowa and that I spent a lot of time on others' farms fortunately gave me enough "street credit" to not make it an issue. Nevertheless, I couldn't help but feel a little resentment toward the question. After all, if I was interviewing, say, grocery store stockers I doubt anyone would care whether or not I was raised in a family of grocery store stockers.

I get it now. Not only is farming based upon a lot of on-the-job training but there are some aspects of food production that one can really only "get" by spending time doing, literally, what farmers do. This makes my respondents' situation all the more interesting because they were not raised in this lifestyle with its associated embodiments. Where and how, then, did those I spoke with learn to do farming if not from this "ladder"?

"It's a hard road, especially at first. You sort of learn as you go, I guess".

Jeff, the author of this quote, described himself as a "gardener turned farmer". For a long time Jeff liked to "piddle in the garden". Then a few years back he tilled up a couple half acre plots in his backyard so he could start selling a variety of vegetables and cut flowers at some of the nearby farmers' markets. The following is an exchange between Jeff and me. We were talking about his experience of entering into farming.

"The first two years were really eye openers".

"How so?", I asked.

"Well, I just never worked so much land before and honestly I wasn't prepared for it. When you're gardening for yourself you can afford to be ignorant a little I guess because if something happens or doesn't work out you can just go to the store or another local grower. But when you're growing stuff for sale, and that's your livelihood, a big chunk of your annual income, you've got to be a little wiser and start managing the land better".

"So how did you learn to manage your land better? Where did you go, who did you talk to, things like that?"

"At first, before I made the leap over into farming, I read a lot of books and did a lot of research on the internet and talked to some friends. That's why I thought I could do it. I was a little overconfident; thought I really knew a lot about all this. [chuckles] Book smarts only get you so far, I'm living proof of that. I mean, I'm I doing good now. Know a lot more; feel good about my knowledge base. But those first couple years; I probably spent just as much time talking to other growers at farmers' markets as customers. I just tried to soak it all up, everything I heard, from how to properly cage my tomatoes to various ways to keep the

raccoons and deer out of my corn. Then I'd take that information home with me
and experiment with it. A lot of trial and error going on before I figured out what
worked best for me and my land".

Jeff made a number of important points during this exchange. For one, he made
the distinction between knowledge that travels well and that which is grounded
within the idiosyncrasies of place and practice. Specifically, Jeff talked about what
he perceived as the limitations of "book smarts" and how that knowledge, which
by definition travels well (after all, it travels around the world in books and via
the internet), needed to be supplemented with "a lot of trial and error". But he also
went beyond this by elevating the latter—more sticky—knowledge, noting it as
essential for figuring out how best to manage his farm. And this process of trial
and error was informed significantly by talking with other growers whom he met
at farmers' markets and putting that knowledge to work.

Research examining the needs of farmers who produce goods for local markets
have noted training and knowledge exchange as important components to success,
especially among those new to the occupation (Braun et al. 2000; Ross 2006, Smith
2006). In many of these studies, "most of the farmers interviewed say that other
farmers have been their most helpful resource" (Ross 2006: 119). This is not to
suggest that conventional farmers do not also look toward other farmers as sources
of knowledge. Yet conventional farmers have long standing social networks that
they can draw upon (Carolan 2006b: 275–9). Indeed, as most individuals climb the
rungs of the agricultural ladder they need not look far to find other farmers who
they can direct questions at. And of course this says nothing about all the other
experts available to commodity farmers—seed and fertilizer dealers, University
Extension personnel, and so forth. Without such built-in interpersonal resources,
respondents described having to be, as quoted below, "a little more creative than
your normal farmer when it comes to finding answers to questions".

> Rebecca: "If you're not a social person I think you're going to have a much
> harder time at it. As someone who's basically a vegetable grower I can't turn
> to my Extension guy or go to the local [Farmers'] Co-op when I have problems
> with my lettuces or artichokes. […] You've got to be a little more creative than
> your normal farmer when it comes to finding answers to questions. Sometimes
> this means just walking up to someone you've never met before and asking them
> about what they do for problems X, Y, and Z".

Michael, one of the respondents who sold eggs for local markets, offered similar
insights about the importance of spaces like farmers' markets for knowledge
exchange among local producers. In his own words:

> You can't read a book about hens and expect to be ready to raise them yourself.
> Come on. Nobody does that. I'm mean, the books don't hurt but you've got to
> get closer to the experience than that. Talking to people who raise hens is a great

start. Going to look at their birds is a great start. I think having chickens will really start picking up once there's a critical mass of people who raise the birds. Because as more people raise hens that means there will be more people to go to for those interested in raising birds. It becomes a virtuous circle.

Going back to the notion of having to be creative (as mentioned by Rebecca) when it comes to establishing conduits through which knowledge—both mobile and sticky—can travel across a social body, I am reminded of the Tour de Coop that occurred in Fort Collins, CO, in the spring of 2009. In an attempt to literally give people a feel for chicken coops, this event involved a bike tour to four established backyard chicken coops within the city of Fort Collins. As discussed in the previous chapter, consumers' distance from livestock and livestock production often exceeds the distance encountered in our relationship to fruit, vegetable, and grain production. While few today are raised on farms, many nevertheless garden, have driven by fields in the countryside, and have seen vegetables and fruit in relatively unprocessed states. Conversely, meat consumers rarely spend much time around livestock or in concentrated animal feeding operations (CAFOs). Similarly, animals are often so processed by the time they hit the shelves at our grocery store that one has to use their imagination if they wish to believe that this "thing" in front of them came from a living, breathing creature. While the knowledge is not entirely substitutable, having grown, for example, a variety of different vegetables in my lifetime I feel comfortable that my gardening skills will help me considerably next spring when I plant a raspberry garden in my backyard. I would not feel as comfortable drawing upon this knowledge base, however, were I planning on building a chicken coop and raising hens. Perhaps Michael was onto something when he argued:

Local growers who raise livestock related commodities in particular need ways to network and interact with others. Most people, unless you happen to be raised on a farm, start off completely clueless on how to raise hens. This is not a skill most people acquire in a normal modern lifestyle.

SSE too was described as an important resource for production oriented knowledge. Discussing the popularity of heirloom tomatoes among local consumers and the difficulty that comes with raising a variety for the first time, Nick explained:

Everyone loves them. I even have a local café calling to see if they can buy directly from me. [...] I want to be able to keep up with demand so I grow more. But being someone who also believes in the important of preserving biodiversity I don't try to grow only one type of heirloom variety. I mix it up. It makes it fun. But when you're trying out a variety for the first time you're taking quite a risk, from a business standpoint, because, not having grown the variety before, you can never be sure just what the plant wants and needs. [...] Talking to folks I met through seed savers [SSE] is huge for me; really important. I don't want to make those rookie mistakes so I talk to others who have. Don't get me wrong, there's still a learning curve to it. But if I didn't have access to the folks at seed savers and the folks I met because of that place I definitely wouldn't be experimenting with heirlooms as much as I have been.

Craig also spoke at length about the importance of SSE as a useful resource for instilling knowledge and skill within potential and current farmers. For Craig, however, it was as much what is not said at SSE as what is said that makes this space epistemologically valuable. In his own words:

Perhaps one of the most valuable things about its [SSE's] gardens is just being able to walk through them. Think about it from my perspective [as someone that sells vegetables at a local farmers' market]. I want to give my customers what they want. I really do. I don't want to tell them what they want. I want them to tell me. […] When I walk the gardens, just when they're starting to produce, I get to see not just theplants but their fruit. It's great. You know, when you buy seed for the first time all you got to go on is the picture on the seed packet, and maybe you'll go online and read about what other people said about the variety. But at seed savers [SSE] you get to hold that tomato in your hand or whatever it is you're looking at growing.

These spaces, whether we are talking about CSA, heritage seed bank, backyard chicken coop, or farmers' market, are more than just places to get food (or seed). They may lack ivy, chalkboards, and lecture halls, but they are without question spaces of learning. Farmers are "schooled" as much within these spaces as the mouths (as detailed in the three previous chapters) they seek to feed.

Building Communities of Practice

Economists talk about the importance of markets. Fair enough, we've got to be able to sell our stuff to consumers. But when you focus entirely on building markets you fail to ask the equally important question. How do we bring farmers—long time farmers, new farmers, and potential farmers—together to make sure we've got people to grow stuff for this market? It's the old chicken and egg thing. Yes, demand's important but let's not forget we also need people who can supply that demand.

This quote came from one of my conversations with Paul. Paul raised and sold eggs for local markets. We were talking about what he saw as the most significant barriers to small-scale, local market oriented food production. Paul was not alone in highlighting what he saw as the dangers of focusing "entirely on building markets" to the exclusion of making sure we have "people who can supply that demand". Rebecca, for example, explained to me how she has "been in contact with people from ISU [Iowa State University] about being involved in partnerships with local grocers and restaurants". About those partnerships she remarked: "That's great, but as someone committed to locally produced, fresh food I also want to figure out ways to get others involved in this type of agriculture too". Jeff echoed this point when he told me:

> Folks like you need to think about how we can increase both supply and demand. But not demand like, how can we increase yields; that mindset has already got us into too much trouble. I'm not talking about coming up with new inputs or new machines, I'm talking about building relationships. And not just between consumers and farmers, which is something things like farmers' markets have been touted as doing, but also between farmers. We need to think of ways to build better relationships between farmers.

Encapsulated in each of these quotes is a nod to the importance of "community" for building local food production capacity. On the subject of barriers encountered by those looking to do small-scale, local market oriented agriculture, Paul talked about the worth of "bringing farmers [...] together", Rebecca mentioned the need for involvement, and Jeff discussed the significance of "building relationships". This reminds me a lot of what is known as "communities of practice".

While conventional learning theory centers largely on the significance of abstract knowledge the concept of communities of practice shines light on the knowledge-creating experience (Lave and Wenger 1991; Wenger 1998; Wenger et al. 2002). Research into these communities reveals the importance they play in the exchange of otherwise "sticky" knowledge—knowledge that can only be acquired through practice. Communities of practice have been shown to create social affordances among its members, which is said to "scaffold knowledge creation in practice" (Brown and Duguid 2001: 203; see also Benkler 2007: 470). The concept of communities of practice is thus a reminder that *practice* must be made an explicit development goal, whether talking about "development" in the conventional sense or in a more limited sense as is the case here (e.g., development of robust local food systems). For "only by first spreading the practice in relation to which the explicit makes sense is the circulation of explicit knowledge worthwhile" (Brown and Duguid 2001: 204).

An early attempt to sketch out the development of these "communities" is provided by Wenger and colleagues (2002). They note how communities of practice often start out as loose networks. Yet, as others have pointed out (Grant 1996: 115; Baalen et al. 2005: 303), even loosely composed networks presuppose *some* network connectivity. This is based upon the empirically documented fact that, ultimately, "the coordination and sharing of knowledge cannot take place without assuming a vast amount of mutual knowledge, mutual beliefs, and mutual assumptions" (Baalen et al. 2005: 303). The lack of this "common ground" (Baalen et al. 2005: 303) can sometimes hold back the formation of strong communities of practice.

Turning our attention back to agriculture, such common ground is built into the traditional agricultural ladder. Individuals born into commodity agriculture are exposed to ways of thinking and doing that are a consequence of the embodiments linked to the agricultural ladder. As I've detailed elsewhere (Carolan 2005), these embodiments come with some shared understandings; a common ground, if you will. Perhaps this is why farmers, upon first meeting me, often ask if I was raised on a farm—a crude measure of establishing whether I share some of this common ground with them. The farmers I spoke to for this research, however, lacking something like an agricultural ladder, had to obtain these "mutual knowledge, mutual beliefs, and mutual assumptions" (Baalen et al. 2005: 303) elsewhere. Earlier I hinted at how individuals acquire some of these basic social competencies that in time will grease exchanges of sticky knowledge. Recall how respondents got their first taste of small-scale, local market oriented agriculture: Lyle first learned about it while attending farmers' markets as a child with his dad; Jeff was initially introduced to this social world during college when he stopped at roadside produce stands and later from farmers' markets and CSA; and Susan's common ground came from friends who were involved in what she called "the local food scene". Perhaps there is a bit of an agricultural ladder hidden in this after all—or, so as not to over-generalize, you might say respondents stepped on some rungs that look an awful lot alike.

Having developed some common ground in turn allowed respondents to further intensify their interactions and knowledge exchanges with those already practicing small-scale, local market oriented agriculture. As Nick explained:

> Once you've developed a rapport with other growers and they know you're serious they're a great resource. [...] I don't think twice about asking them questions; learning from them. Some of them have been growing for so long that they've got an answer for you even before you're done describing your problem.

Debbie talked about a similar intensification of networks when she described her evolution from a local food consumer to a local food producer. When developing the initial common ground her interactions were, as she describes, "awkward". Yet, with time, as Debbie developed social, linguistic, and embodied competencies, she was able to make those interactions feel "natural". In her own words:

> I'll admit it, when I was new to this my relationship with other farmers was on the awkward side. I really wanted to learn about it but I was just too green [new to farming]; didn't have the language down yet, didn't know terms or I used them incorrectly. [...] Now it's not a problem. My interactions with other growers are much more natural-feeling now.

Local Knowledge Spillover

Research into communities of practice has often centered on the innovative capacity that exists between organizations, such as between firms or between firms and universities (see, for example, Ponds et al. 2010; Powell et al. 1996). While most often used to discuss issues related to innovation, knowledge exchange, and building shared understandings at the intra- and inter-firm levels, the concept nicely encapsulates a lot of different themes that emerged out of the interviews. If communities can and do instill practice, knowledge, skills, and understandings when it comes to tuning bodies for farming—whether we are talking about large-scale commodity agriculture or small-scale, local market oriented agriculture—then perhaps we might be able to find other parallels between the worlds of business and local food systems. While not often referred to in discussions about local food systems (though there are exceptions [such as Turner 2010]), I would like to talk briefly about what is known as knowledge spillover.

Knowledge spillovers describe the leaking of information between organizations. Academic knowledge spillovers are most frequently discussed in the literature, where the knowledge generated within university units spills over into private industry. This explains the emergence of science parks near major research universities, such as in Boston and San Francisco. The literature identifies a couple of mechanisms through which this spillover occurs. One is labor mobility. As individuals move from one organization to the next they take with them not only any explicit knowledge but also knowledge that is more sticky, such as any skills and practices that allow them to more efficiently put that explicit knowledge to work (Ponds et al. 2010). Another spillover mechanism is informal networks that emerge as a result of spatial proximity. Research suggests that local, informal linkages offer more open channels for information transfer than the weaker network ties that typically stretch across private industry (Owen-Smith and Powell 2004: 10).

Take the phenomenon of labor mobility. I spoke to a couple of individuals who attended two, three, and sometimes even four different farmers' markets a week. This level of farmers' market participation aided in the expansion of a "community" of practice for it broadened those informal linkages. Spillover was also helped along by a "soft infrastructure" (Turner in press), which involves things like email, formal associations (such as being a member of SSE), electronic forums (such as SSE's forum <http://forums.seedsavers.org/>), and blogs (such as SSE's blog <http://blog.seedsavers.org/>). This helps explain why innovations quickly spread through the community (see also Hinrichs et al. 2004). As Ted explained:

> Someone might come up with a new way of displaying their produce, maybe bagging it a certain way, and if it ends up being a hit with the shoppers then next week you'll see others trying something similar. I remember last year a guy came to [farmers'] market with a super sweet variety of sweetcorn. [...] His was always the first to go. Now you've got three or four people selling that stuff. [...]

It's not like we're secret about new ideas. Why bother? We're not blind. We can see what works and what doesn't.

On the point of labor mobility within local food systems and its affect on the spatiality of knowledge exchange, Debbie offered the following insights:

When a corn and soybean producer goes to market, that involves a trip to the local [farmers'] Co-op. One trip there, one trip back. They pull their load in, dump it, and leave. When I take my harvest to market, I go a lot of places and I come into contact with a lot of people. We see each other every week, over and over again for about four or five months. And for me, I see a lot of different people because I go to multiple markets. [...] Not only do I get to meet a lot of great people but it allows me to talk shop with others who love to work the earth too. I learn a little here, give a little there.

These sentiments were echoed by Joyce:

Once you're recognized by other growers as someone that's serious about this then suddenly the lips start getting loose. I mean that in a good way. [...] There are some really knowledgeable folks out there. Especially those that do this for a living. [...] Those folks go to a market almost every day, which in itself is a great way to learn all the secrets on how to be successful in this business.

Spillover was also discussed in the context of what I earlier called soft infrastructure, which exists in the form of formal associations and connectivities made possible by electronic and print media. My point in mentioning this is not to suddenly stray away from the embodied forms of knowing and learning that I've spent the better part of this book empirically detailing. I have said repeatedly that representational knowledge does not exist separate from forms of knowledge that cannot be as easily codified. I would like to spend some time now detailing the ways that this epistemological reality revealed itself during the interviews.

A concern has been raised by some scholars that fixing communities of practice to spatial proximity unintentionally reinforces problematic dualisms (Amin and Cohendet 1999; Bathelt et al. 2004). For example, on the point that tacit knowledge can only be exchanged via local networks, Faulconbridge (2006: 518) worries that this creates "a powerful set of discourses [...] that create a misleading dualism between tacit and explicit knowledge and local and global geographies, respectively". Is it possible to have such things as regional, state, or perhaps even global communities of practice (Faulconbridge 2006: 518–19; Wenger et al. 2002: 25)? Wenger and colleagues (2002: 25) seem to think so:

[M]any communities start among people who work at the same place or live nearby. But colocation is not necessary. Many communities of practice are distributed over wide areas. Some communities meet regularly [...]. Others are connected primarily by e-mail and phone and may meet only once or twice a year. What allows members to share knowledge is not the choice of a specific form of communication (face-to-face as opposed to Web-based, for instance) but the existence of a shared practice—a common set of situations, problems and perspective.

Yet what *type* of knowledge is being shared once communities become so distanced that words substitute for face-to-face, tactile encounters? Among those I spoke to at least, sticky knowledge was *not* what was being exchanged through this soft infrastructure. Take the following exchange between me and Nick, a member of SSE.

"Their [SSE] mailings [that come with a membership] are interesting. Sometimes you get a helpful tip or a recipe. I occasionally check their website, like their blog. Actually I've gotten a couple of interesting pointers from their forum. It's also accessed through their website. [...] They give me some ideas but I wouldn't say I learn anything from them".

"Could you explain that? What do you mean you don't learn anything? How is getting ideas not the same as learning?", I asked.

"Well, I guess, it's not, it's not that I'm not learning anything from them, I just mean that until I've tried them out I can't really say what works. The website and things like that point me in certain directions but I guess I don't consider it 'learning' until I put those ideas to work because what works for someone in a different climate and soil type might not work for me".

When talking about the dislocation of communities of practice we must be careful not to conflate spatial reach with social depth (Morgan 2004: 4). For Nick, things like email, the internet, and newsletters are great vehicles for mobile knowledge. Yet mobile knowledge is *not* the same as practical, embodied knowledge. In fact, Nick was uncomfortable labeling this more mobile knowledge as "knowledge", and preferred instead the term "ideas". Only through practice, by putting those ideas to work, can Nick identify what knowledge ought to be learned and what information can be discarded.

Nick was not alone in making this distinction between knowledge acquired through more disembodied media and knowledge acquired through the rough and tumble world of practice. Michael talked at great length about how "the local chicken scene, while growing, remains pretty small". This caused him to look outward, especially to the internet. Now he "has quite an extensive e-community with others who also have an interest in raising chickens". But, like Nick, Michael was uncomfortable equating knowledge acquired from his "e-community" with that learned through what he later called "trial and error". In his own words:

You can't take anything you learn through a website or blog posting at face value; even if they're from trusted sources. When it comes down to it, you really don't know anything until you've taken it through your own little backyard laboratory. It's still ultimately a discovery process. There is no substitute for the actual experience seeing what works for you.

I think the distinction between "networks of practice" and "communities of practice" is useful here. Unlike communities of practice—where skills, understandings, and other forms of sticky knowledge are transferred over face-to-face, tactile experiences—networks of practice involve links that "are mostly indirect (e.g., databases, newsletters, info bulletins) and members [...] do not interact directly [...] and produce little knowledge" (Baalen et al. 2005: 302–3). This distinction is important because it gives us a language to talk about spatially proximate interactions—communities of practice—without casting respondents as insular or unconnected to larger (regional, national, international) networks. Michael made this point best when he said:

I've got a group of friends, fellow chicken enthusiasts, we'll get together, I don't know, about once a month, have a beer and talk chickens. Sometimes we'll get together at one of our places. Someone's always trying something new or has a new gadget. This way we can see what others are doing and learn from each other. That's my number one go-to resource. [...] I'm connected to a broader network of birders, I don't live on an island, but this group of friends is my most important resource. I wouldn't be where I am today without them.

The Local: An Analytic Category Pointing to Something Real

There is a burgeoning literature on the subject of "the politics of scale" (see, for example, Agnew 1994, 1997; Born and Purcell 2006; Brenner 1997, 1999, 2001; Howitt 1998; Marston 2000; Smith 1992). This literature "offers the a priori conclusion that there is nothing inherent about scale", arguing that "scale and scalar configurations are not an independent variable that can *cause* outcomes, rather they are a *strategy* used by political groups to pursue a particular agenda" (Brown and Purcell 2005: 608). In other words, scale is a social construction (Marston 2000). Yet the more time I spend studying embodied knowledge the more I question those who claim that "scale has no ontological nature" (Brown and Purcell 2005: 609).

Within agro-food studies this argument has been used to conceptually disassemble those that make a case for local food. Proponents of local food are said to have fallen into the "local trap" (Born and Purcell 2006); something they would do well to climb out of given that they are promoting a thing with no ontological basis. Yet what exactly are the trappings of the local food literature? The architects of the local trap argument are not entirely clear in their answer to this question. If they mean that "the local trap is the assumption that local is inherently good"

(Born and Purcell 2006: 195), then I agree with them. Agro-food scholars have been saying for over 10 years that the terms "local", "ecologically sound", and "socially just" ought not to be conflated (Allen and Guthman 2006; DuPuis and Goodman 2005; Hinrichs et al. 1998; Hinrichs 2000, 2003; Winter 2003). Jeff reminded me of this when he said: "As farmers we still need to make money and sometimes that causes us to do things that might not be all that ecologically sound—like driving my old gas burning pickup truck to markets 50 miles away each way". It is when the local trap is conceptualized as the fetishization of a thing that doesn't exist—namely, scale—that I begin scratching my head.

Framers of the local trap argument claim that "[n]o matter what its scale, the outcomes produced by a food system are contextual: they depend on the actors and agendas that are empowered by the particular social relations in a given food system" (Born and Purcell 2006: 195–6). I agree with the first part of the above sentence—that these outcomes are contextual. The latter half, however, does not go far enough, reflecting a too disembodied understanding of "contextual".

The research for this book has led me to conclude that there really is, ontologically speaking, something real about spatially proximate relations. Do I think scalar definitions are political? Yes. As *categories*, are the terms "local" or "global" socially constructed? Of course. Is there nothing inherent about scale? The evidence does not allow me to answer this question in the affirmative. Michael's earlier quote, for example, where he explains that his distant network of birders is no substitute for the tactile, face-to-face encounters of spatially proximate chicken enthusiasts, implies that there is something inherent about local relations. Yet he was not alone in making this distinction.

> Craig: "There is no substitute for being there. Rural Northeastern Iowa is not exactly a center of activity for local food. It's taken me a while to build up my network [of small-scale local-market-oriented farmers]. You spend hours poring through books and the internet but I've learned that the best way to learn about farming is by going to other farms and seeing what they're doing. Like I said, that's why I like the gardens at seed savers".

> Susan: "You'd never try to learn to fly by reading books and sitting in chat rooms with pilots. Same with farming. There's this intangible quality to learning and working directly with others".

Decades of research back up my assertion that local relationships matter for handing over—pun intended—certain forms of knowledge. The sociologist of science Harry Collins (1974, 1992), for example, famously describes the difficulties encountered when attempts were made to replicate a TEA (Transverse Electrical discharge in gas at Atmospheric pressure) laser. According to Collins' account, constructing the first TEA laser involved tremendous trial and error. Yet once the first properly functioning laser was built, others still found it surprisingly difficult to replicate the technology. Attempts to replicate this technology ended with devices that either failed to work at all or did so only sporadically. No matter how methodically teams followed the instructions of the inventors they could not produce an adequately working model. Yet, when these teams added members who had produced working models they were eventually able to construct a functional TEA laser. Collins notes that no functional TEA lasers where developed using only published sources. Embodied, tacit knowledge proved essential for making an operational TEA laser.

Latour notes something similar in his ethnographic study of French engineering students in Côte d'Ivoire. The students were given blueprints of automobile engines and instructed to repair the broken engines that lay before them. Never having worked on an automobile engine before, the students had a difficult time relating the information in the text to the engines. Not until they came into contact with others in possession of the requisite embodied knowledge were they able to create an operating engine (Laet and Raven 1989). An ethnographic study of Xerox service technicians explained how the technical manuals were of little use compared to the immense skill (and sometimes improvisation) acquired from engaging daily with other service technicians and the various tools and machines of their trade (Orr 1996). In sum, as Laet (2000: 163) explains, "words only reveal the things for those who have *made* the things into the words—or who have made the things and the words simultaneously—with their own hands, eyes, and brains first".

> I helped cut asparagus today. Nice day, upper 70s with very little humidity. Working with two other people. One of the guys had never cut asparagus before. I never realized just how much knowledge I had of asparagus until today. He was having a hard time knowing what to cut and what to leave to turn to seed. It's not just about cutting asparagus of a certain length. There's not really a rule you can follow. The asparagus just has to feel a certain way. We were trying to tell this guy that—because he was cutting stuff that wasn't really all that big but was too hard to eat (or at least it wouldn't have tasted the best). But how can you tell someone how ready-to-cut asparagus should feel? Finally [Stanley] walked over to him, took his hand that had the knife in it, and held it against a couple asparagus sticks. "Feel that; that's too hard", he explained as they cut into the first plant. Then they moved to the plant just left of the first one. Cutting into it [Stanley] proclaimed, "Now there's your perfect asparagus; see how the knife just goes right through".

Cutting asparagus is admittedly not at the top of most farmers' lists of must have knowledge. But that's not the reason why I reproduced this self-reflexive account of my time harvesting this vegetable. This exchange could have just as easily been about Stanley teaching someone about how healthy soil should feel between one's fingers, or how to evaluate soil based upon the tractor's response during cultivation, or how to assess a hen's health by watching, listening, and holding it. Learning to farm involves close encounters like those between Stanley and the asparagus cutting novice. Just another reason why spaces like those detailed in this book are important: they help make these close encounters possible.

The Pedagogies of Agriculture

I was careful earlier when talking about sticky knowledge not to say that one model of food production presupposes more than the other. I've been around enough agricultural production methods to know all rely significantly upon embodied knowledge. And I am still not ready to concede that one requires, *in toto*, more embodied knowledge than the other. It's clear, however, that sticky knowledge makes up a significant portion of the expertise needed to successfully *do* small-scale local market oriented food production. And I am convinced that a gradual expansion of this form of production can only occur with the help of those spaces discussed in this book.

This brings me to the following observation made by Susan: "You've got some places in this country that are complete food deserts. No local food anywhere and no one either willing or with the knowledge to start things going. That's a problem. How's anyone going to learn if they've got no one to learn from?" Building on Susan's point, while important not to over generalize when making distinctions between "local" and "global" (recognizing that this distinction is itself a generalization) small-scale, local models of food production *by definition* presuppose a strong grasp of local conditions. And I'm not just talking about growing conditions; though, as others have argued (see e.g., Kloppenburg 1991; McAfee 2003), conventional agriculture works pretty hard to smooth away local conditions allowing standardized production practices to be put to work over large expanses of space and time. Lest we forget: many *producers* for local markets are also *marketers* for local markets, which means they also have to know what their consumers what to taste, smell, see, and feel by way of locally raised foods. Conventional growers, in contrast, need only look as far as what processors— grain crushing facilities, meat packers, etc.—want (which is often explicitly communicated through contracts).

Debbie believed the contrast between what small-scale, local market oriented farmers (as both producers *and* marketers) and conventional farmers (as producers) need to know has not been adequately recognized by those who study food systems. And until it is, we will continue to miss an opportunity to adequately train the next generation of growers for alternative markets. In her own words:

> "Land grant universities do a great job educating people on how to farm conventionally, use the latest technologies, and participate in the global market. I think the skill set you need to be a manager of a CSA, however, is entirely different. Since every city's different, with a different customer base, different tastes, likes, and dislikes, I just don't see you getting that sort of specific knowledge from a course that supposed to teach you about CSAs, as if they're all the same. Conventional ag. might like this cookie cutter approach to educating

the next generation of farmers but I don't think that will fly if we're talking about trying to teach people about how to run a successful CSA".

To which I asked: "So what do you suggest needs to be done to best educate people on how to run a CSA or be a grower for a farmers' market?"

"I think it requires a different approach to education entirely. Something like an apprenticeship; something like that. Everyone I know learned from others. That's not to say that universities can't be involved. But the approach needs to be different. Less classroom time. I don't know; they need more time doing and less time thinking".

More time doing and less time thinking: that's an interesting choice of words. It suggests that conventional classroom pedagogies are more centered around a disembodied understanding of cognition, where knowledge is something one *acquires* not something one *does*. And there is some truth to this caricature of the educational system (Goodman 2008; Gunter 2006). Debbie is reminding us that there are multiple ways to get an education when it comes to wanting to be trained to become a small-scale, local market oriented producer. While a formal education, such as a four-year college degree, has value for those looking to learn how to farm, it's important not to miss, according to Debbie, the pedagogical value of *doing* farming too.

When it has come to training the next generation of conventional farmers, the issue of tacit knowledge has always largely taken care of itself. Growing up on a farm and climbing "up" the agricultural ladder virtually guarantees the transference of sticky knowledge to these apprentice farmers. My conventional farming friends who briefly left the farm to go off to college went there to learn about the more explicit side of food production—to learn about things like accounting, food policy, and agricultural economics. Don't get me wrong, I know from personal experience that Land Grant Universities give students a lot of hands-on experience too. I wish not to appear as though I am suggesting that universities only deal in the currency of explicit knowledge. My point, rather, is that in training conventional farmers Land Grant Universities never had to worry as much about the more "sticky" side of what it means to *do* production agriculture, thanks to the agricultural ladder. When it comes to small-scale, local market oriented food production, given the absence of this "ladder", we can no longer make this assumption. Where is this sticky knowledge going to come from if alternative producers don't follow the conventional path assumed by the agricultural ladder? This question needs to become a more regular part of the food policy debate, especially if we are serious about creating small-scale, local market food systems. Tentatively speaking, my research suggests that the local food systems *themselves* might be a valuable source of this knowledge.

I shall conclude by briefly reaching beyond the thirteen individuals who inform this chapter. This discussion of the pedagogical value of *doing* small-scale food production places the rise of urban gardens, a phenomenon that has repeatedly made the local and national news (at least in the US) in recent months (see e.g., Beras 2010), in a new and hopeful light. While it saddens me *why* urban gardens are experiencing a renaissance in the US—all arrows point to the global depression we find ourselves in—I am encouraged by what this could mean for the future of small-scale, local market-oriented food production. As more people *do* urban agriculture more people, as we've learned from this chapter, will have the opportunity to learn (and get excited) about this form of food production, which, in turn, will hopefully lead to more people *doing* urban agriculture, and so forth.

There's also an important structural component to the recent growth of urban gardens that could bode well for future growth in this area. City ordinances erected during the twentieth century were terribly unfriendly to anyone looking to raise food within the boundaries of an urban corridor. The recent growing interest in producing food within city limits has forced city planners to revisit their code, which many of them are. These policy changes will give the inhabitants of cities greater freedom to create local food networks—to allow not only gardens but also bee hives, chickens, pigs, goats, and the like. I am hopeful that these structural changes will be of some embodied consequence as more people are allowed to *do*, and become a part of (rather than a part from), food production.

Chapter 7
Steps to an Ecology of Social Change

Questioning the transformational potential of spaces like those discussed in earlier chapters, the editors for the book *Hungry for Profit*, an edited collection of essays critical of corporate agriculture, ask "whether this pathway is really a solution to the problems or rather something that will produce only a minor irritant to corporate dominance of the food system" (Magdoff et al. 2000: 188). They further maintain that "a complete transformation of the agriculture and food system, it might be argued, requires a complete transformation of society" (p. 188). There is also a fair amount of suspicion among agro-food scholars of consumerist-driven accounts seeking to challenge corporate controlled food (see, for example, Guthman 2007). Those critical of consumer-led social change are right to highlight the problems—such as their classist and racial/ethnic undertones (Guthman 2008b; Slocum 2008) and their non-reflexiveness (DuPuis and Goodman 2005; Winter 2003)—of these "pocket book" approaches. Contrary to the neoliberal narrative taught to us in grade school, being a good consumer is not the same as being a good citizen (Dean 1999). And more important still: voting with money only works if you have some. Let us not forget: being able to make a choice presupposes that you are in a position to have a choice and possess the resources to carry that decision out. Even among those with money it is not evenly distributed, creating still further asymmetries if we were to buy into the "democracy = free market" argument.

Nor do I entirely understand why we've elevated ethical food *consumption* as the preferred pathway to "citizenship'. This leads Goodman and colleagues (2010: 1783) to ask, "Why are poor peasants in Guatemala not more citizenly for *growing* fair trade coffee? What about organic farmers in Lincolnshire or Ohio or Sardinia?" Then there is the question of non-consumption. What is to become of those that do not consume, either because they are poor or because they've made a lifestyle choice to live that way? Equating consumption with citizenship virtually guarantees their status of something less than a citizen. In the eyes of a market democracy they are not as important as those whose affluence allows them to over consume, which is like stuffing the ballet box.

To be sure, it's psychologically comforting to conflate being a "good" citizen with buying the "right" things. The idea that ecological sustainability and social justice can be attained by merely making the correct purchasing decisions, which we've been conditioned to equate with buying products with terms like "green", "organic", "fair trade", "local", and "carbon neutral" printed on them, is an empowering thought. The problem (or at least one of them) with this approach is that it focuses attention on the *food* part of the food system at the expense of further masking its structural and organizational components—in other words, the *system* part. Forgetting about the system reinforces "the idea that social change is simply a matter of individual will rather than something that must be organized and struggled over in collectivities" (Allen and Guthman 2006: 412).

Whether we work to challenge and transform existing food systems, as suggested by the editors of *Hungry for Profit*, or take a more piecemeal position and create alternative food pathways that may someday supplant the old, as suggested by other agro-food scholars (see, for example, Hassanein 2003; Henderson 2000), the requisite of either position is the same. Both require knowledge of alternatives as well as individuals *wanting* to act. What are the (embodied) pathways that make people think and behave as they do? To put it another way—and to evoke language from Allen and Guthman's aforementioned quote—how do we make people feel like they *ought* to organize, struggle, and act collectively? Calls for collective action absent of research into the questions of *why* and *how* people think and act as they do come at the expense of sounding hollow and insincere.

I don't mean to sound harsh. It is the social scientist's job to understand the constraints and opportunities of social life. It is how we presently seem to understand these phenomena that I have a problem with. Constraints in the agro-food literature are usually defined as external to the individual, as evidenced by the prevalence of political economic frameworks. When conducted in absence of a lived approach, like that attempted here, political economy frameworks give the appearance that if/when external constraints are lifted the "proper" behavior will result. In truth, I know of no scholars studying food systems who would agree with this highly structural caricature of human behavior. But in the absence of more sensual accounts describing how and why we act as we do, agro-food literature tends to convey just such a picture of the social world. As I hope to have communicated through the previous chapters, understandings, knowledge, and thus behaviors are, at least in part, effects of embodied relations. If we hope to understand social change, and perhaps even play a part in shaping its trajectory, we will need to expand our understanding of "constraints" to include assemblages of the body.

To alleviate concern that this book is unreflexive in its attention to structural concerns—as I have accused political economy approaches as being unreflexive toward lived concerns—this chapter more explicitly interweaves the structural with the lived. To do this I step back from the case studies and attempt to glean some policy recommendations from the earlier chapters. Recognizing that the process of tuning bodies to alternative food systems requires assemblages of the body, this chapter makes explicit the reality that those assemblages include phenomena like the state, markets, and capital.

Tuning from the "Top"

Previous chapters describe a virtuous relationship between the spaces described and the embodied knowledge that flows out of these relationships. The former was described as feeding the latter, which in turn reinforces the former, and so forth. I heard the story repeatedly: these spaces made Global Food problematic in the eyes (and bodies) of respondents. Once tuned to the food coming from these alternative foodscapes interviewees actively sought these spaces.

Yet what happens when these spaces are new? What are we to do in those places where these alternative food spaces have yet to form and where potential consumers (and producers) are not yet tuned to them? Such systems, as mentioned, appear to have a self-reinforcing quality. What is to be done, however, in those communities where individuals only know "food" through Global Food; where alternative tunings are not readily available?

Sociologists of technology talk about how non-mainstream technological artifacts sometimes require, if they are ever to mature, an incubation space. This refers to a space where novel technologies are shielded from market forces (Carolan 2010b: 48; Geels 2007: 1414). These spaces allow for the formation of things like learning, scale, and network economies, as individuals literally get the chance to *do* the technology in question (Unruh 2000: 220–26). The field of business management has a similar term: business incubation. Though used to explain various phenomena, the understanding that interests me here reflects reasoning similar to that of incubation space: namely, small businesses, if viewed as socially and economically desirable, ought to be shielded from market forces (see e.g., Biggs 2002). The rationale is that such protection will help ensure their survival, especially early on before having developed customer loyalty and/or a niche market not properly served by large firms.

Famers' markets have been singled out by agro-food scholars as an example of one type of business in need of incubation. Given the relatively low cost of entering into such a business they have been viewed as potential launching pads for entrepreneurialism (Gillespie et al. 2008; Guthrie et al. 2006). To be sure, the entry cost into small-scale, local food systems can be low. Yet that is not always the case. For example, while land is plentiful and, relatively speaking, inexpensive in many rural areas, the same does not hold when talking about the New Yorks and Hong Kongs of the world. I worry that the costs of market entry are so great in these spaces that without some degree of non-market intervention the sensual experiences described in this book will continue to elude large swaths of the world's population who reside in urban centers.

Ultimately, it's a moral decision. If we as a society think these spaces—and the food they provide—are important then we are going to have work actively (and, yes, collectively) to make these spaces and experiences possible. The market alone will not supply them; not as long as the alternative—namely, Global Food—continues to enjoy all the privileges that go along with market dominance. Lest we forget, Global Food has its own spaces of incubation, through not only a billion dollars of government subsidies daily (Peterson 2009) but also thanks to policies and regulations that allow many of its true "costs" to be externalized onto society writ large (Carolan 2011).

It has been over a decade since James Scott's (1999) book *Seeing Like a State* was published. The book centers on failed cases of large-scale authoritarian plans in a variety of fields. Scott argues that monolithic social plans often misfire because they replace complex interdependencies with one-size-fits-all schematic visions. He also argues that successful designs for social organization depend upon viewing tacit, practical knowledge as being as important as explicit, abstract knowledge. As a whole, Scott is highly critical of organizational forms designed to increase societal well being that disregard the values and desires of its subjects. Scott provides a compelling example of what it means to see like a state. But do we *feel* like a state?

When talking about agriculture it's hard to take the analytic scalpel and make a clean cut between the worlds of production and politics. The lines between, say, the Corn Growers Association, American Soybean Association, Monsanto, and USDA are blurred as people, organizational goals, and values are shared (see e.g., Nestle 2003: 99–110). The "state" in this case thus refers to much more than elected officials or government bureaucrats making policy decisions. The agrofood literature is clear on this point: the state of Global Food is quite literally the food of the state (see e.g., Clapp and Fuchs 2009; Weis 2007; Winders 2009).

Between 1995 and 2005, the US government paid approximately $164.7 billion in agricultural subsidies, with a third of that ($51.3 billion) going to subsidize corn production (Philpott 2007). Yet the state's involved—and long has been—in tuning us to Global Food in still other ways. Take the National School Lunch Program. Created in 1946, the National School Lunch Program was advertised to offer free meals to poor children and subsidize healthy lunches for all school children. It was justified with nutritional science. Yet, as argued by Susan Levine (2008: 39), "it bore only slight resemblance to the goals of nutrition scientists and home economists. The program was, in its goals, structure, and administration, more a subsidy for agriculture than a nutrition program for children". This program has been more than just an agriculture subsidy. It has also literally helped tuned bodies to Global Food.

Or take US food aid. In 1954, the Agricultural Trade Development and Assistance Act, also known as Public Law 480 (PL 480), was passed. The PL480 focused on export subsidies to create new markets for US agricultural commodities in the developing world through the use of international food aid (Winders 2009: 82). The law authorized the exchange of excess agricultural commodities for foreign currency. Initially, the program was used for "surplus removal" but later was redefined as "self help" for countries suffering from shortages of fats and oils (Goldberg 1968: 126). But the "help" these programs offer do nothing (or at least very little) to sustain local agricultural systems. Rather than a tool for surplus disposal, export promotion, and creating geopolitical leverage to benefit privileged domestic interests (Barrett and Maxwell 2005: 26–35), food aid policies could provide real aid, where long-term poverty and hunger reductions and the sustaining of traditional grips are primary goals.

The USAID Food Aid budget in 2005 was US$1.6 billion. Of that, only 40 percent (US$654 million) went to paying for food. The rest was spent on overland transportation (US$141 million), ocean shipping (US$341 million), transportation and storage in destination country (US$410 million), and administrative costs (US$81 million) (Dugger 2005). Perhaps there are more optimal ways to spend this money; ways more interested in supporting local knowledge, practices and lived experiences of food production and consumption. As opposed to making people feel (and see) like a state why not support feelings, practices, and *doings* of food that are in tune with the bodies and needs of those receiving the aid?

In 2003, the USDA allocated US$333 million for nutrition education (Schoonover and Muller 2006: 11). Compare this to the US$10–15 billion spent annually on food and beverage advertising aimed at children. According to an article in *Advertising Age*, PepsiCo earned $43 million in worldwide sales in 2008. Some of its product-specific advertising expenditures in 2008 for what's known as "measured media"—expenditures funneled through advertising agencies— include US$162 million for Gatorade, US$145 million for Pepsi Cola, US$27 million for Tostitos, US$14 million for Doritos, and US$11 million for Fritos. Not included in these figures are money Pepsi spends on things like lobbying, supporting the American Beverage Association's efforts to fight soda taxes, a Pepsi-funded obesity study at Yale, or marketing to children and adults throughout the developing world (Nestle 2010).

All this money is having the desired effect. US caloric consumption increased from 2,234 calories per person per day in 1970 to 2,757 calories in 2003. Studies have consistently found a positive correlation between number of television hours watched by children and frequency of requests, purchases, and servings eaten of advertised food (see e.g., Claney-Hepburn and Hickey 1974; Frances et al. 2003; Taras et al. 1989). According to the *Guinness Book of Records*, the McDonald arches are more widely recognized in the world than the Christian Cross. Children consume approximately 167 additional calories for each hour of TV watched (Wiecha et al. 2006). A child's risk for obesity increases by 6 percent for every hour of TV watched on average each day (Robinson et al. 2001). In another study, preschool children reported that food in McDonald's wrappers tasted better than food in plain wrappers even though *the same* food item was used in each treatment (Robinson et al. 2007).

And where is the state in all of this? It has essentially incentivized this push to tune our bodies to nutritionally empty calories with the US Internal Tax Revenue Code of 1986. This legislative change to US tax code gives *tax breaks* to companies for advertising and marketing unhealthy food to all bodies, including kids.

Enabling Ethics

Andrew Berry (2004: 199–203) has written on what he calls the "ethical assemblage". According to Berry (2004: 200), "[r]ather than examine ethics as an abstract set of principles, such an approach would focus on the specificity of the discursive and non-discursive assemblages that are expected to generate ethical forms of conduct". To give an example of the ethical assemblage, Berry (2004: 200) points to the jury:

> Their ethical capacity to judge a crime on the basis of the evidence that is presented before them depends on the spatial and procedural organization of the court. Although the law assumes that all citizens have, in principle, the capacity to make disinterested judgments, this capacity is not intrinsic to the citizen but is an effect of the ethical assemblage of which he or she comes to play a part.

Agro-food scholars refer to a similar ethical effect when they talk about the power of, for example, labels (see e.g., Goodman 2004: 903–8; Raynolds 2002: 409–12). Labels attempt to expand the consumer's ethical capacity to judge food. Those judgments, like the capacity to judge a crime, are not intrinsic to the individual. They are, rather, a relational effect. Fair trade and organic labels, though but a proxy for something else, still "seek to 'lengthen' across the spaces of consumption" (Goodman and Goodman 2001: 111) an individual's being-in-the-world. Labels work to bring certain aspects of the food system more clearly into focus (Goodman 2004: 906).

Can we think of ways to support alternative food networks and enable people to become part of what Barry calls an "ethical assemblage"? I have already talked about how the state could play a role in protecting these spaces by way of creating an incubation space, which would allow certain competencies to form while being shielded from so called "market" forces. But what about consumers? If ethical action is a *process*, which is not only Barry's point but also something I've been arguing throughout this book, then there is something unjust about the fact that social position allows some people to act more ethically than others. Having money—and time for those who choose to raise some of their own food— increases one's ability to have the type of lived experiences described in previous chapters. Finding ways to enable people, especially those living on the margins, to become a part of (rather than apart from) these assemblages should be something we actively strive for. Allowing things like "food stamps" to be used at farmers' markets is a good first step. And, more recently, accepting food debit cards—since food stamps in the US are now paperless—in outdoor farmers' markets is yet another step in the right direction. Yet access to these spaces, as well as to whole foods more generally, remains a problem for those of lower socio-economic status (Allen and Wilson 2008). Clearly, more work needs to be done to make these ethical assemblages available to all, including those who can least afford them.

"Ethicists", as Peterson (2009: 36) argues, "need to pose the sociological question of what kind of selves and relations people need in order to hold each other accountable, to help each other live by our own highest values and be our own best selves". The subject of ethics, I realize, has come up repeatedly throughout this book. The literature on ethics so utterly misses its sociological underbelly that I feel compelled to return to the subject again and again. Individuals, on the whole, do not communicatively (in the Habermasian sense) arrive at their *oughts*. Rather, these ethical convictions—at least those that move us—come from lived experience. This is a profoundly consequential finding.

Ethicists, from Plato to St. Augustine, Kant and Niebuhr, presume that *oughts* precede acts. Morality has long been viewed as a "top-down affair" (Peterson 2009: 128), in that we deduce what to value and then universally apply those *oughts* to concrete situations. The fact that few people are able to lead a purely ethical life, following this model, has been taken by many ethicists as evidence that so called modern life is somehow corrupting or immoral (see e.g., D'Souza 2005; Niebuhr 1934). They never bothered to take this finding as evidence that their own understanding of ethics is what's problematic and that "people do not find or create moral values in a vacuum" (Peterson 2009: 128). I think I've provided compelling evidence in the preceding chapters that oughts are not generated in a vacuum.

And sometimes, like in the case of our being tuned to Global Food, those *oughts* have come about intentionally. Chapter 2 describes a variety of ways in which political and economic forces have worked to make us feel a certain way towards Global Food. In light of this intentionality, and the interests involved in making us "choose" industrial foods, I think we are justified to ask the state to play a role in helping us value small-scale, local market oriented foods; to *do*, in other words, what was done to us by Global Food, with the goal of making industrial food out of tune.

Anna Peterson (2009), in her book *Everyday Ethics and Social Change*, talks about how attitudes and values sometimes follow even those practices that were undertaken without initial willingness. She gives the example of the 1954 US Supreme Court decision *Brown v. Board of Education of Topeka*, the landmark court ruling that struck down the infamous "separate but equal" rule made famous in the *Plessy v. Ferguson* US Supreme Courting ruling from 1899. As Peterson explains (2009: 132):

> Had the Court waited until most white southerners approved of integrated schools, my children might attend segregated schools today. As it stands, while school desegregation certainly did not end racism, it has had a significant effect on the lives and values of both white and black southerners. While southerners had to act as though black people were equal, even though most of them did not believe this in their hearts or minds in 1954, they were not acting morally, by Kantian or Lutheran standards. [...] *Brown* generated major shifts in values that probably would not have occurred had institutions and practices not changed first.

She goes on to suggest that "an environmental *Brown v. Board of Education* could help bring our practices into accord with our expressed values", in addition to the added benefit, given how *oughts* tend to follow practice, of generating "environmental values as well as positive practical results" (Peterson 2009: 132). I am not sure what the food equivalent to an "environmental *Brown v. Board of Education*" would look like, in part because Peterson is terribly short on specifics when it comes to explaining her own thoughts on the subject. But I appreciate her broader point: that if we think a particular way of doing something is fundamentally unsustainable then perhaps we shouldn't wait for "the market" to respond.

Hard core libertarians in the US still bristle at the fact that civil rights legislation was passed in the 1960s, believing that the market—the modern day equivalent of the agora where people talk with their wallets—would have eventually taken care of all forms of overt discrimination. The truth is, however, that the market *hadn't* created the needed changes in how we treated and thought about one another. Those changes in attitudes and behavior only came after—and arguably in part *because of*—the court's *Brown* ruling.

There is no question that the current food system produces food that we simply cannot afford (Carolan 2011). One study, focusing on the effects of agricultural production in the US on natural resources, wildlife, biodiversity, and human health, estimated that the externalized costs of farming alone were somewhere between $5.7 to $16.9 billion annually (Tegtmeier and Duffy 2004: 1). Some well placed "carrots" (e.g., tax incentives, consumer subsidies, etc.) and "sticks" (e.g., more strict environmental regulations) are needed to help shape behaviors of farmers, processors, retail outlets, and consumers. Such action by the state has the potential to generate shifts in values, which would, in turn, self-reinforce those practices and create long lasting behavioral change in the direction of sustainability.

Concluding Thoughts

In the end, it all comes back to building relationships that *do* what we want them to do when it comes to building a sustainable system of food provisioning. Global Food might produce "cheap" food but that type of food comes at great expense to the environment, community, knowledge, and types of sensual experience. Before we can talk about a "better" food system we need to agree on what a "good" food system should do. I, for one, think our food system should do more than just produce food cheaply, especially if its cheapness is a product of bad accounting, where most costs are externalized. Discussing what a food system ought to do— rather than busying ourselves with higher level debates like is slow or fast, small-scale or large-scale, local or global "better"—helps avoid the problematic traps that have sidelined the food debate in recent decades. Doing this allows us to take down the "facile dichotomies between fast and slow, reflexive and compulsive, fat and thin, and, hence, good and bad eaters, to show where there is slippage and instability in these categories, in addition to a troubling politics of class and gender" (Guthman 2003: 45).

It is simply not as black and white as the dichotomies would lead us to believe. Probyn (2001) makes this point beautifully when discussing McDonald's. Eating at McDonald's for many is an expression of care; of caring for one's family and providing for one's children. More than that, Probyn notes, McDonald's provides a convivial family space, which is more than can be said for many spaces where families eat meals these days (during those rare instance when families get together to eat). All this "caring" occurs, however, at the expense of care for the environment, animals, and those people behind the Happy Meal. At McDonald's, "'[t]wo worlds collide' in terms of a vision of care, when, on the one hand, McDonald's stitches us all together through our stomachs, and on the other, a politics that directly equates the desire for burgers with the destruction of the rain forest, and the exploitation of workers and children" (Probyn 2001: 36). Julie Guthman (2003: 56) sums the point up nicely about how categories like slow (or fast) and organic (or conventional) tell us nothing about what a particular food system *is*:

> Fast food is often pitched to healthy eaters (e.g. Subway®'s advertising campaign suggesting you can lose weight and cut fat by eating fast food) and slow food is often made tasty by slavish uses of salt and butter. […] Little is it considered that organic production depends on the same systems of marginalized labor as does fast food. Or that organic salad mix led the way in convenience packaging, and is often grown out of place and out of season. Or that fast food serves women who work outside the home who are then blamed for depending on it to manage family and work. Or that slow food presumes a tremendous amount of unpaid feminized labor.

Building relations that value things like unpaid labor, ecological and social systems, and animal welfare will require an approach that is comfortable walking between the worlds of "inside" symbolic meanings and "outside" political economic structures (recognizing, ultimately, that the two worlds are really one). A decentered approach like that employed in this book allows for such a "conversation" between worlds to occur. Hopefully it will also increase the dialogue in academia between those looking "up" (e.g., political economy approach) and those looking "down" (e.g., interpretive approach) when examining subjects related to food and agriculture.

I have trained this approach on two CSAs, a heritage seed bank, and about two dozen individuals whose backyards are populated by chickens. I leave it to others to direct it to not only other spaces but also other spaces in a non-US context, which would allow for valuable cross cultural comparisons to take place. Though always nice to conclude an academic work by generalizing from the findings I realize such a move, in this case, would be premature. I do think I have shown, however, strong evidence in support of the hypothesis mentioned in the first chapter, where I posit that relationalities underlie all (food) knowledges, making those connectivities inherently political. We can now focus our energies on detailing the various ways that food production, distribution, and consumption are embodied and political. Our understanding of food and the oughts, sentiments, and feelings we attach to the artifacts and processes that feed us are entangled in our "being with" food. Realizing this radicalizes our politics. It reveals what can only be described as an embodied food politics.

References

Abram, David. 1997. *The Spell of the Sensuous*. New York: Vintage Books.

Acampora, Ralph. 2005. Zoos and Eyes: Contesting Captivity and Seeking Successor Practices, *Society and Animals* 13: 69–88.

Adam, Barbara. 1999. Industrial Food for Thought: Timescapes of Risk, *Environmental Values* 8(2): 19–38.

Adi, Bongo. 2005. The Moral Economy and Prospects of Accumulation in Sub-Sahara Africa, *West Africa Review* 7, http://www.westafricareview.com/issue7/adi.html, last accessed June 10, 2010.

Adorno, Theodor. 1984. *Aesthetic Theory*. New York: Routledge.

Agnew, John. 1994. The Territorial Trap: The Geographical Assumptions of International Relations Theory, *Review of International Political Economy* 1(1): 53–80.

Agnew, John. 1997. The Dramaturgy of Horizons: Geographical Scale in the "Reconstruction of Italy" by the New Italian Political Parties, 1992–95, *Political Geography* 16(2): 99–121.

Allen, Patricia. 2007. *Together at the Table: Suitability and Sustenance in the American Agrifood System*. University Park, PA: Pennsylvania State University.

Allen, Patricia and Alice Wilson. 2008. Agrifood Inequalities: Globalization and Localization, *Development* 51(4): 534–40.

Allen, Patricia and Julie Guthman. 2006. From "Old School" to "Farm-to-School": Neoliberalization from the Ground Up, *Agriculture and Human Values* 23(4): 401–15.

Allen, Patricia, Margaret Fitzsimmons, Michael Goodman, and Keith Warner. 2003. Shifting Plates in the Agrofood Landscape: The Tectonics of Alternative Food Initiatives in California. *Journal of Rural Studies* 19: 61–75.

Allen, Patricia and Martin Kovach. 2000. The Capitalist Composition of Organic: The Potential of Markets in Fulfilling Promise of Organic Agriculture, *Agriculture and Human Values* 17(3): 221–32.

Allman, William. 1995. *Stone Age Present: How Evolution Has Shaped Modern Life*. New York: Simon and Schuster.

Amin, Ash and Patrick Cohendet. 1999. Learning and Adaptation in Decentralized Business Networks, *Environment and Planning D: Society and Space* 17: 87–104.

Anderson, Benedict. 1993 (1983). *Imagined Communities: Reflections on the Origins and Spread of Nationalism*. London: Verso.

Anderson, Virginia. 2004. *Creatures of Empire*. New York: Oxford University Press.

Arendt, Hannah. 1958. *The Human Condition*. Chicago, IL: University of Chicago Press.

Armitage, Kevin. 2007. "The Child Is Born a Naturalist": Nature Study, Woodcraft Indians, and the Theory of Recapitulation, *Journal of Gilded Age and Progressive Era* 6(1) <http://www.historycooperative.org/cgi?bin/justtop.cgi?act=justtop&url=http://www.historycooperative.org/journals/jga/6.1/armitage.html>, last accessed September 30, 2010.

Atkinson, David. 2007. Kitsch Geographies and the Everyday Spaces of Cultural Memory, *Environment and Planning A* 39: 521–40.

Avise, John. 2004. *The Hope, Hype, and Reality of Genetic Engineering*. New York: Oxford University Press.

Baalen, Peter van, Jacqueline Bloemhof-Ruwaard, and Eric Heck. 2005. Knowledge Sharing in an Emerging Network of Practice: The Role of a Knowledge Portal, *European Management Journal* 23(3): 300–14.

Bagdonis, Jessica, Claire Hinrichs and Kai Schafft. 2009. The Emergence and Framing of Farm-to-School Initiatives: Civic Engagement, Health and Local Agriculture, *Agriculture and Human Values* 26: 207–19.

Bailey, L.H. 1903. *The Nature-Study Idea: Being an Interpretation of the New School Movement to Put the Child in Sympathy with Nature*. New York: Doubleday.

Barnett, Clive and David Land. 2007. Geographies of Generosity: Beyond the Moral Turn, *Geoforum* 38: 1065–75.

Barrett, Christopher and Daniel Maxwell. 2005. *Food Aid after Fifty Years: Recasting its Role*. New York: Routledge.

Bates, Diane and Thomas Rudel. 2004. Climbing the "Agricultural Ladder": Social Mobility and Motivations for Migration in an Ecudaorian Colonist Community, *Rural Sociology* 69: 59–75.

Bateson, Gregory. 1973. *Steps to an Ecology of the Mind*. London: Granada.

Bathelt, Harold, Andersand Malbberg, and Peter Maskell. 2004. Clusters and Knowledge: Local Buzz, Global Pipelines and the Process of Knowledge Creation, *Progress in Human Geography* 28(1): 31–56.

Batsell W. Robert, Alan Brown, Matthew Ansfield, and Gayla Paschall. 2002. "You Will Eat All of That!": A Retrospective Analysis of Forced Consumption Episodes, *Appetite* 38: 211–19.

Bauman, Zygmunt. 1991. *Modernity and the Holocaust*. Oxford: Polity.

Bauman, Zygmunt. 1993. *Postmodern Ethics*. New York: Blackwell.

Beck, Ulrich. 1992. *Risk Society: Towards a New Modernity*. London: Sage.

Beck, Ulrich. 1999. *World Risk Society*. New York: Wiley-Blackwell.

Becker, Howard and Anselm Strauss. 1956. Careers, Personality and Adult Socialization, *American Journal of Sociology* 62(3): 253–63.

Becket, J.W. 1969. A Career Ladder for Farm Workers, *Farm Labor Developments* December: 7–11.

Bekoff, Marc. 2002. *Minding Animals: Awareness, Emotions and Heart*. Oxford: Oxford University Press.

Bell, Michael Mayerfeld. 1994. *Childerley: Nature and Morality in a Country Village*. Chicago, IL: University of Chicago Press.

Bell, Michael Mayerfeld. 1997. The Ghosts of Place, *Theory and Society* 26(6): 813–36.

Bellah, Robert, Richard Madsen, William M. Sullivan, Ann Swidler, and Steven M. Tipton. 1985. *Habits of the Heart*. Berkeley, CA: University of California Press.

Benkler, Yochai. 2007. *The Wealth of Networks: How Social Production Transforms Markets and Freedom*. New Haven, CT: Yale University Press.

Beras, Erika. 2010. Cities Grapple with Rise of Urban Agriculture, National Public Radio Program, Aired on April 24th, 2010. Transcript available on-line at http://www.publicbroadcasting.net/weku/news.newsmain/article/0/1/1642598/Central.and.Eastern.Kentucky/Cities.Grapple.with.Rise.of.Urban.Agriculture, last accessed August 1, 2010.

Berger, John. 1980. *About Looking*. London: Writers and Readers.

Bergson, Henri. 2007 (1911). *Matter and Memory*. New York: Candler Press.

Berlan, Jean-Pierre, Jean-Pierre Bertrand, and Laurence Lebas. 1977. The Growth of the American "Soybean Complex", *European Review of Agricultural Economics* 4(4): 395–416.

Berndsen, Mariette and Joop van der Pligt. 2004. Ambivalence Towards Meat, *Appetite* 42: 71–8.

Berry, Andrew. 2004. Ethical Capitalism, in *Global Governmentality*, edited by Wendy Larner and William Walters, pp. 195–211, New York: Routledge.

Berry, Wendell. 1990. *What Are People For?* New York: North Point Press.

Berry, Wendell. 2007. *Conversations with Wendell Berry*. Jackson, MS: University of Mississippi Press.

Bickerstaff, Karen and Gorden Walker. 2003. The Place(s) of Matter: Matter Out of Place—Public Understandings of Air Pollution, *Progress in Human Geography* 27: 45–67.

Biggs, Tyler. 2002. Is Small Beautiful and Worthy of Subsidy? International Finance Corporation (IFC). Washington, DC, http://citeseerx.ist.psu.edu/viewdoc/download?doi=10.1.1.113.385&rep=rep1&type=pdf, last accessed February 23, 2010.

Birch, Leann. 1980. Effects of Peer Models' Food Choices and Eating Behaviors on Preschoolers' Food Preferences, *Child Development* 51(2): 489–96.

Blay-Palmer, Alison. 2008. *Food Fears: From Industrial to Sustainable Food Systems*. Burlington, VT: Ashgate.

Bondi, Liz. 2005. Making Connections and Thinking through Emotions: Between Geography and Psychotherapy, *Transactions of the Institute of British Geographers* 30: 433–48.

Born, Branden and Mark Purcell. 2006. Avoiding the Local Trap: Scale and Food Systems in Planning Research, *Journal of Planning Education and Research* 26(2): 195–207.

Bourdieu, Pierre. 1984. *Distinction: A Social Critique of the Judgment of Taste.* Cambridge, MA: Harvard University Press.

Bourdieu, Pierre. 1995. *Outline of a Theory of Practice.* New York: Cambridge University Press.

Braidotti, Rosi. 2006. *Transpositions: On Nomadic Ethics.* Malden, MA: Polity.

Braun, Ann, Graham Thiele, and Maria Fernandez. 2000. Farmer Field Schools and Local Agricultural Research Communities: Complementary Platforms for Integrated Decision Making in Sustainable Agriculture, Agricultural Research Extension Network, UK Department for International Development, London, Paper No. 105, 1–16.

Brenner, Neil. 1997. State Territorial Restructuring and the Production of Spatial Scale: Urban and Regional Planning in the FRG, 1960–1989, *Political Geography* 16: 273–306.

Brenner, Neil. 1999. Globalization as Reterritorialization: The Re-scaling of Urban Governance in the European Union, *Urban Studies* 36(3): 431–51.

Brenner, Neil. 2001. The Limits to Scale? Methodological Reflections on Scalar Structuration, *Progress in Human Geography* 25(4): 591–614.

Brown, Cheryl and Stacy Miller. 2008. The Impacts of Local Markets: A Review of Research on Farmers Markets and Community Supported Agriculture, *American Journal of Agricultural Economics* 90(5): 1296–302.

Brown, Christopher and Mark Purcell. 2005. There's Nothing Inherent About Scale: Political Ecology, the Local Trap, and the Politics of Development in the Brazilian Amazon, *Geoforum* 36: 607–24.

Brown, John and Paul Duguid. 2001. Knowledge and Organization: A Social-Practice Perspective, *Organization Science* 12(2): 198–213.

Brown, Kristen. 2006. *Nietzsche and Embodiment: Discerning Bodies and Non-Dualism.* Albany, NY: SUNY Press.

Bruegel, Martin. 2002. How the French Learned to Eat Canned Food, 1809–1930, in *Food Nations: Selling Taste in Consumer Societies*, edited by Warren Belasco and Philip Scranton, pp. 113–30, New York: Routledge.

Bubandt, Nils. 1998. The Odour of Things: Smell and the Cultural Elaboration of Disgust in Eastern Indonesia, *Ethnos* 63, pp. 48–80.

Bulliet Richard. 2005. *Hunters, Herders, and Hamburgers.* New York: Columbia University Press.

Burnett, Charles. 1991. The Superiority of Taste, *Journal of the Warburg and Courtauld Institutes* 54: 230–38.

Burnham, George. 1855. *The History of the Hen Fever: A Humorous Record.* Boston: Hobart and Robbins.

Buttel, Fredrick, Olaf Larson, and Gilbert Gillespie. 1990. *The Sociology of Agriculture.* Santa Barbara, CA: Greenwood Press.

Caldwell, Melissa. 2004. Domesticating the French Fry: McDonald's and Consumerism in Moscow, *Journal of Consumer Culture* 4(1): 5–26.

Campbell, Brian. 1985. Uncertainty as Symbolic Action in Disputes Among Experts, *Social Studies of Science* 15(3): 429–53.

Carolan, Michael. 2005. Barriers to the Adoption of Sustainable Agriculture on Rented Land: An Examination of Contesting Social Fields, *Rural Sociology* 70(3): 387–413.

Carolan, Michael. 2006a. Sustainable Agriculture, Science, and the Co-Production of "Expert" Knowledge: The Value of Interactional Expertise, *Local Environment: The International Journal of Justice and Sustainability* 11: 421–31.

Carolan, Michael S. 2006b. Social Change and the Adoption and Adaptation of Knowledge Claims: Whose Truth Do you Trust in regard to Sustainable Agriculture?, *Agriculture and Human Values* 23: 270–85.

Carolan, Michael S. 2006c. Do You See What I See? Examining the Epistemic Barriers to Sustainable Agriculture, *Rural Sociology* 71(2): 232–60.

Carolan, Michael S. 2007. Introducing the Concept of Tactile Space: Creating Lasting Social and Environmental Commitments, *Geoforum* 38: 1264–75.

Carolan, Michael S. 2008. The More-Than- Representational Knowledge/s of Countryside: How We Think as Bodies, *Sociologia Ruralis* 48(4): 408–22.

Carolan, Michael S. 2009a. I Do Therefore There Is: Enlivening Socio-Environmental Theory, *Environmental Politics* 18(1): 1–17.

Carolan, Michael S. 2009b. Ethanol versus Gasoline: The Contestation and Closure of a Socio-technical System in the USA, *Social Studies of Science* 39(3): 421–48.

Carolan, Michael S. 2010a. *Decentering Biotechnology: Assemblages Built and Assemblages Masked*. Farnham, UK: Ashgate.

Carolan, Michael S. 2010b. *A Sociological Look at Biofuels: Understanding the Past/ Prospects for the Future*. Hauppauge, NY: Nova Science Publishers.

Carolan, Michael S. 2011. *The Real Cost of Cheap Food*. London: Earthscan.

Carpenter, Edmund. 1973. *Eskimo Realities*. Austin, TX: Holt, Rinehart and Winston.

Cartwright, John. 2000. *Evolution and Human Behavior: Darwinian Perspectives on Human Nature*. Cambridge, MA: MIT Press.

Charters, Stephen. 2006. *Wine and Society: The Social and Cultural Context of a Drink*. Woburn, MA: Butterworth-Heinemann.

Claney-Hepburn, Katherine, Anthony Hickey, and Gayle Nevill. 1974. Children's Behavior Responses to TV Food Advertisements, *Journal of Nutrition Education* 6: 93–96.

Clapp, Jennifer and Doris Fuchs. 2009. Agrifood Corporations, Global Governance, and Sustainability: A Framework for Analysis, in *Corporate Power in Global Agrifood Governance*, edited by Jennifer Clapp and Doris Fuchs, pp. 1–26, Cambridge, MA: MIT Press.

Clark, Andy. 1999a. Where Brain, Body, and World Collide, *Journal of Cognitive Systems Research* 1: 5–17.

Clark, Andy. 1999b. *Being There: Putting Brain, Body and the World Together Again*. Cambridge, MA: MIT Press.

Clark, Andy. 2008. *Supersizing the Mind: Embodiment, Action and Cognitive Extension*. New York: Oxford University Press.

Coats, C. David. 1989. *Old MacDonald's Factory Farm*. New York: Continuum.

Cochrane, Willard. 1993. *The Development of American Agriculture: A Historical Analysis*. Minneapolis, MN: University of Minnesota Press.

Cohen, Benjamin. 2009. The Once and Future Georgic: Agricultural Practice, Environmental Knowledge, and the Place for an Ethic of Experience, *Agriculture and Human Values* 26(3): 153–65.

Collins, Harry. 1974. The TEA Set: Tacit Knowledge and Scientific Networks, *Science Studies* 4: 165–86.

Collins, Harry. 1992 (1985). *Changing Order: Replication and Induction in Scientific Practice*. Chicago, IL: University of Chicago Press.

Connerton, Paul. 1989. *How Societies Remember*. Cambridge: Cambridge University Press.

Connolly, William. 2003. *Neuropolitics: Thinking, Culture, Speed*. Minneapolis, MN: University of Minnesota Press.

Cook, Harold. 1996. Physicians and Natural History, in *Cultures of Natural History*, edited by Nicholas Jardine, Jim Secord, and Emma Spary, pp. 91–105, New York: Cambridge University Press.

Coombes, Brad and Hugh Campbell. 2002. Dependent Reproduction of Alternative Modes of Agriculture: Organic Farming in New Zealand, *Sociologia Ruralis* 38(2): 127–45.

Cresswell, Tim. 1996. *In Place/Out of Place: Geography, Ideology, and Transgression*. Minneapolis, MN: University of Minnesota Press.

Damerow, Gail. 2002. *Barnyard in Your Backyard: A Beginners Guide to Raising Chickens, Ducks, Geese, Rabbits, Goats, Sheep, and Cattle*. North Adams, MA: Storey Publishing.

Dean, Mitchell. 1999. *Governmentality: Power and Rule in Modern Society*. Thousand Oaks, CA: Sage.

De Laet, Marianne and Annemarie Mol. 2000. The Zimbabwe Bush Pump: Mechanics of a Fluid Technology, *Social Studies of Science* 30: 225–63.

Deleuze, Gilles. 1995. *Difference and Repetition*. New York: Colombia University Press.

Delind, Laura. 2006. Of Bodies, Place and Culture: Re-situating Local Food, *Journal of Agricultural and Environmental Ethics* 19(1): 121–46.

Delind, Laura and Anne Ferguson. 1999. Is This a Women's Movement? The Relationship of Gender to Community Supported Agriculture in Michigan, *Human Organization* 58(2): 190–200.

Deloria, Philip. 1999. *Playing Indian*. New Haven, CT: Yale University Press.

Denzin, Normand. 1989. *Interpretive Interactionism*. Newbury Park, CA: Sage.

Dewey, John. 1897. My Pedagogical Creed, *School Journal* 54(3): 77–80.

Diner, Hasia. 2001. *Hungering for America: Italian, Irish and Jewish Foodways in the Age of Migration*. Cambridge, MA: Harvard University Press.

Douglas, Mary. 1966. *Purity and Danger*. London: Routledge and Kegan Paul.

Douglas, Mary. 1982. *In the Active Voice*. New York: Routledge and Kegan Paul.

DuBois, Christine. 2008. Social Context and Diet: Changing Soy Production and Consumption in the United States, in *The World of Soy*, edited by Christine DuBois, Chee-Beng Tan, and Sidney Mintz, pp. 208–33, Urbana and Chicago, IL: University of Illinois.

Duffy, R., A. Fearne, and Y. Healing. 2005. Reconnecting in the UK Food Chain: Bridging the Communication Gap Between Food Producers and Consumers, *British Food Journal* 107(1): 17–33.

Dugger, Celia. 2005. Africa Food for Africa's Starving is Road Blocked in Congress, *New York Times* October 12, http://www.nytimes.com/2005/10/12/ international/africa/12memo.html?ex=1286769600&en=0de1afa6dd7990e7& ei=5090&partner=rssuserland&emc=rss, last accessed February 19, 2010.

DuPuis, E. Melanie and David Goodman 2005. Should We Go "Home" to Eat?: Toward a Reflexive Politics of Localism, *Journal of Rural Studies* 21(3): 359–71.

Durkheim, Emile. 2001 (1912). *The Elementary Forms of Religious Life*. New York: Oxford University Press.

D'Souza, Albert. 2005. *Christen Ethics and Moral Values*. New Delhi: Mittal Publications.

Eden, Sally, Christopher Bear, and Gordon Walker. 2008. Mucky Carrots and Other Proxies: Problematising the Knowledge-fix for Sustainable and Ethical Consumption, *Geoforum* 39: 1044–57.

Elias, Norbert. 1996 (1989). *The Germans: Power Struggles and the Development of Habitus in the Nineteenth and Twentieth Centuries*. New York: Colombia University Press.

Elias, Norbert. 1999. *The Germans*, Preface by Eric Dunning and Stephen Mennell. New York: Columbia University Press.

Elias, Norbert. 2000 (1939). *The Civilizing Process*. New York: Wiley.

Ensminger, Audrey. 1993. *Foods and Nutrition Encyclopedia*. Boca Raton, FL: CRC Press.

Evans-Pritchard, E. 1940. *The Nuer*. Oxford: Clarendon Press.

Falk, Pasi. 1994. *The Consuming Body*. Thousand Oaks, CA: Sage.

Fantasia, Rick. 1995. Fast Food in France, *Theory and Society* 24(2): 201–43.

FAO (Food and Agriculture Organization). 2006. *Livestock's Long Shadow: Environmental Issues and Options*. Rome, Italy: FAO.

Faulconbridge, James. 2006. Stretching Tacit Knowledge Beyond a Local Fix? Global Spaces of Learning in Advertising Professional Service Firms, *Journal of Economic Geography* 6: 517–40.

Feld, Steven. 1996. Waterfalls of Song, in *Senses of Place*, edited by Steven Feld and Keith Basso, pp. 91–135, Santa Fe, NM: School of American Research Press.

Feld, Steven. 1984. *Sound and Sentiment: Birds, Weeping, Poetics, and Song in Kaluli Expression*. Philadelphia, PA: University of Pennsylvania Press.

Ferguson, Kennan. 2004. I Heart My Dog, *Political Theory* 32(3): 373–95.

Ferguson, Priscilla. 2004. *Accounting for Taste: The Triumph of French Cuisine*. Chicago, IL: University of Chicago Press.

Feyerabend, Paul. 2000. *The Conquest of Abundance: A Tale of Abstraction vs. the Richness of Being*. Chicago, IL: University of Chicago Press.

Findlen, Paula. 1994. *Possessing Nature: Museums, Collecting, and Scientific Culture in Early Modern Italy*. Berkeley, CA: University of California Press.

Findlen, Paula. 1996. Courting Nature, in *Cultures of Natural History*, edited by Nicholas Jardine, Jim Secord, and Emma Spary, pp. 57–75, Cambridge: Cambridge University Press.

Fine, Ben. 2004. Debating Production-Consumption Linkages in Food Studies, *Sociologia Ruralis* 44: 332–42.

Fitzgerald, Gerard and Gabriella Petrick. 2008. In Good Taste: Rethinking American History with Our Palates, *The Journal of American History* 95(2): 392–404.

Foster, E.M. 2002. E.M. Foster on Prunes and English Food, in *Choice Cuts: A Savory Selection of Food Writing from Around the World and Throughout History*, edited by Mark Kurlansky, pp. 404–6, New York: Ballantine Books.

Foucault, Michel. 1980. *Power/Knowledge: Select Interviews and Other Writings, 1972–1977*. New York: Pantheon Books.

Foucault, Michel. 1983. *This is Not a Pipe*. Berkeley, CA: University of California Press.

Foucault, Michel. 2002. *Archaeology of Knowledge*. New York: Routledge.

Frances, Lori, Yoona Lee, and Leann Birch. 2003. Parental Weight Status and Girl's Television Viewing, Snacking, and Body Mass Indexes, *Obesity* 11: 143–51.

Freidberg, Susanne. 2003: Editorial: Not all Sweetness and Light: New Cultural Geographies of Food, *Social and Cultural Geography* 4: 3–6.

Freidberg, Susanne. 2004. *French Beans and Food Scares: Culture and Commerce in an Anxious Age*. New York: Oxford University Press.

Freidberg, Susanne. 2009. *Fresh: A Perishable History*. Cambridge, MA: Harvard University Press.

Friedmann, Harriet. 2007. Scaling Up: Bringing Public Institutions and Food Service Corporations into the Project for a Local, Sustainable Food System in Ontario, *Agriculture and Human Values* 24(3): 389–98.

Garson, Marjorie. 2007. *Moral Taste: Aesthetics, Subjectivity and Social Power in the Nineteenth-Century Novel*. Toronto: Toronto University Press.

Geels, Frank. 2007. Analysing the Breakthrough of Rock 'n' Roll (1930–1970): Multi-Regime Interaction and Reconfiguration in the Multi-Level Perspective, *Technological Forecasting and Social Change* 74: 1411–31.

Geertz, Clifford. 1996. *After the Fact: Two Countries, Four Decades, One Anthropologist*. Cambridge, MA: Harvard University Press.

Gething, Anna. 2010. Menstrual Metamorphosis and "The Foreign Country of Femaleness", in *Rites of Passage: In Postcolonial Women's Writing*, edited by Pauline Dodgon-Katiyo and Gina Wisker, pp. 267–82, New York: Rodopi.

Gibson, James. 1986. *The Ecological Approach to Visual Perception.* New York: Lawrence Erlbaum.

Gibson, Sarah. 2007. Traveling, Dwelling, and Eating Cultures, *Space and Culture* 10(1): 4–21.

Giddens, Anthony. 1990. *The Consequences of Modernity*. Cambridge: Polity.

Giddens, Anthony. 1991. *Modernity and Self-Identity: Self and Society in the Late Modern Age*. Cambridge: Polity.

Gillespie, Gilbert, Duncan Hilchey, Clare Hinrichs, and Gail Feenstra. 2008. Farmers' Markets as Keystones in Rebuilding Local and Regional Food Systems, in *Remaking the North American Food System: Strategies for Sustainability*, edited by Clare Hinrichs, and Thomas A. Lyson, pp. 65–83, Lincoln, NE: University of Nebraska Press.

Gillette, Mrs. F.L. 1889. *White House Cookbook*. Chicago, Philadelphia, Stockton, CA: L.P. Miller and Company.

Goldberg, Ray. 1968. *Agribusiness Coordination: A Systems Approach to the Wheat, Soybean, and Florida Orange Economies*. Boston, MA: Harvard School of Business.

Goodman, David. 2001. Ontology Matters: The Relational Materiality of Nature and Agro-food Studies, *Sociologia Ruralis* 41(2): 182–200.

Goodman, David. 2003. Editorial: The Quality 'Turn' and Alternative Food Practices: Reflections and Agenda, *Journal of Rural Studies* 44: 1, 3–16.

Goodman, David. 2004. Rural Europe Redux: Reflections on Alternative Agro-Food Networks and Paradigm Change, *Sociologia Ruralis* 44: 3–16.

Goodman, David and E. Melanie Dupuis. 2002. Knowing Food and Growing Food: Beyond the Production-Consumption Debate in the Sociology of Agriculture, *Sociologia Ruralis* 42(1): 6–23.

Goodman, David and Michael Goodman. 2001. Sustaining Foods: Organic Consumption and the Socio-Ecological Imaginary, in *Exploring Sustainable Consumption: Environmental Policy and the Social Sciences*, edited by Maurie Cohen and Joseph Murphy, pp. 97–119, Oxford: Elsevier Science.

Goodman, Michael. 2004. Reading Fair Trade: Political Ecological Imaginary and the Moral Economy of Fair Trade Foods, *Political Geography* 23: 891–915.

Goodman, Michael. 2008. "Did Ronald McDonald also Tend to Scare you as a Child?": Working to Emplace Consumption, Commodities and Citizen Students in a Large Classroom Setting, *Journal of Geography in Higher Education* 32(3): 365–86.

Goodman, Michael. 2010a. The Mirror of Consumption: Celebritization, Developmental Consumption and the Shifting Cultural Politics of Fair Trade, *Geoforum* 41: 104–16.

Goodman, Michael. 2010b. Towards Visceral Entanglements: Knowing and Growing the Economic Geographies of Food, in *The Sage Companion of Economic Geography*, eds. R. Lee, A. Leyshon, L. McDowell and P. Sunley, in press, London: Sage.

Goodman, Michael, Damien Maye, and Lewis Holloway. 2010. Ethical Foodscapes? Premises, Promises and Possibilities, *Environment and Planning A* 42(8): 1782–96.

Granovetter, Mark. 1985. Economic Action and Social Structure: The Problem of Embeddedness, *American Journal of Sociology* 91: 481–510.

Grant, Robert. 1996. Toward a Knowledge-Based Theory of the Firm, *Strategic Management Journal* 17: 109–22.

Gunter, Pete. 2006. Whitehead and Environmental Education, in *A Different Three Rs for Education: Reason, Relationality, Rhythm*, edited by George Allen and Malcolm Evans, pp. 75–86, New York: Rodopi.

Guthman, Julie. 2002. Commodified Meanings and Meaningful Commodities: Re-thinking Production and Consumption Links Through the Organic System of Provision, *Sociologia Ruralis* 42: 295–311.

Guthman Julie. 2003. Fast Food/Organic Food: Reflexive Tastes and the Making of "Yuppie Chow", *Social and Cultural Geography* 4(1): 45–58.

Guthman, Julie. 2007. Commentary on Teaching Food: Why I am Fed Up with Pollan et al., *Agriculture and Human Values* 24: 261–4.

Guthman, Julie. 2008a. If They Only Knew: Color Blindess and Universalism in California Alternative Food Institutions, *The Professional Geographer* 60(3): 387–97.

Guthman, Julie. 2008b. Bringing Good Food to Others: Investigating the Subjects of Alternative Food Practice, *Social and Cultural Geography* 15: 431–47.

Guthrie, John, Anna Guthrie, Rob Lawson, and Alan Cameron. 2006. Farmers' Markets: The Small Business Counter-revolution in Food Production and Retailing, *British Food Journal* 108(7): 560–73.

Habermas, Jürgen. 1984. *The Theory of Communicative Action*. Cambridge: Polity.

Hahn, E. 1967. *Animal Gardens.* Garden City: Double Day.

Halbwachs, Maurice. 1992 (1941). *On Collective Memory*. Chicago, IL: Chicago University Press.

Hancharick, Amber and Nancy Kiernan. 2008. Improving Agricultural Profitability Through an Income Opportunities for Rural Areas Program, *Journal of Extension* 46(5) at http://www.joe.org/joe/2008october/a3.php, last accessed August 1, 2010.

Haraway, Donna. 1991. *Simons, Cyborgs and Women: The Reinvention of Nature*. London: Routledge.

Haraway, Donna. 2003. *The Companion Species Manifesto: Dogs, People, and Significant Otherness*. Chicago, IL: Prickly Paradigm Press.

Harris, Marshall. 1950. A New Agricultural Ladder, *Land Economics* 26(3): 258–67.

Harris, Marvin. 1986. *Good to Eat: Riddles of Food and Culture*. London: Allen and Unwin.

Harrison, Paul. 2000. Making Sense: Embodiment and the Sensibilities of the Everyday, *Environment and Planning D: Society and Space* 18: 497–517.

Haruf, Kent. 1999. *Plainsong*. New York: Alfred A. Knopf, Inc.

Hassanein, Neva. 1999. *Changing the Way America Farms: Knowledge and Community in the Sustainable Agriculture Movement*. Lincoln, NE: University of Nebraska Press.

Hassanein, Neva. 2003. Practicing Food Democracy: A Pragmatic Politics of Transformation, *Journal of Rural Studies* 19: 77–86.

Hassanein, Neva and Jack Kloppenburg. 1995. Where the Grass Grows Again: Knowledge Exchange in the Sustainable Agriculture Movement, *Rural Sociology* 60(4): 731–40.

Havelock, Eric. 1986. *Reflections on Orality and Literacy from Antiquity to the Present*. New Haven, CT: Yale University Press.

Hayes-Conroy, Allison and Jessica Hayes-Conroy. 2008. Taking Back Taste: Feminism, Food, and Visceral Politics, *Gender, Place and Culture* 15(5): 461–73.

Heidegger, Martin. 2000 (1926). *Being and Time*. Hoboken, NJ: Wiley.

Henderson, Elizabeth. 2000. Rebuilding Local Food Systems from the Grass-Roots Up, in Magdoff, Fred, John Foster, and Frederick Buttel, (eds), *Hungry for Profit: The Agribusiness Threat to Farmers, Food, and the Environment*, pp. 175–88, New York: Monthly Review Press.

Henderson, Elizabeth and Robyn Van En. 2007. *Sharing the Harvest: A Citizen's Guide to Community Supported Agriculture*. White River Jct, VT: Chelsea Green Publishing.

Hendrickson, Mary and William Heffernan. 2002. Concentration of Agricultural Markets, Unpublished paper, Department of Rural Sociology, University of Missouri, Columbia, MO, http://www.foodcircles.missouri.edu/CRJanuary02.pdf, last accessed June 1, 2010.

Hinchliffe, Steve, Matthew Kearnes, Monica Degen, and Sarah Whatmore. 2005. Urban Wild Things: A Cosmopolitical Experiment, *Environment and Planning D: Society and Space* 23: 643–58.

Hinrichs, Clare. 2000. Embeddedness and Local Food Systems: Notes on Two Types of Direct Agricultural Markets, *Journal of Rural Studies* 16(3): 295–303.

Hinrichs, Clare. 2003. The Practice and Politics of Food System Localization, *Journal of Rural Studies* 19: 33–45.

Hinrichs, Clare, Gilbert Gillespie, and Gail Feenstra. 2004. Social Learning and Innovation at Retail Farmers' Markets, *Rural Sociology* 69(1): 31–58.

Hinrichs, Clare, Jack Kloppenburg, George Stevenson, Sharon Lezberg, John Hendrickson, and Kathryn DeMaster. 1998. *Moving Beyond Global and Local*. United States Department of Agriculture, Regional Research Project NE-185 working statement, October 2, <http://www.ces.ncsu.edu/depts/sociology/ne185/global.html>, last accessed November 5, 2009.

Hipple, Eric von. 1994. "Sticky Information" and the Locus of Problem Solving: Implications for Innovation, *Management Science* 40(4): 429–39.

Hodge, Clifton. 1902. *Nature, Study and Life*. Boston, MA: Ginn and Company.

Holstein, James and Jaber Gubrium. 2000. *Constructing the Life Course*. Landam, MD: AltaMira.

Honore, Carl. 2004. *In Praise of Slow*. Orion: London.

Howitt, Richard. 1998. Scale as Relation: Musical Metaphors of Geographical Scale, *Area* 30(1): 49–58.

Hudson, Ian and Mark Hudson. 2003. Removing the Veil? Commodity Fetishism, Fair Trade, and the Environment, *Organization and Environment* 16(4): 413–30.

Hughes, Thomas. 1969. Technological Momentum in History: Hydrogenation in Germany 1898–1933, *Past and Present* 44(1): 106–32.

Humphrey, Nicholas. 1995. Introduction: Histories, *Social Research* 62(3): 477–79.

Hurst, Daniel. 2010. Growers Go Bananas Over Waste, *Brisbane Times* January 7, http://www.brisbanetimes.com.au/business/growers-go-bananas-over-waste-20100106-lu7q.html, last accessed July 30, 2010.

Hutchings, J.B. 1977. The Importance of Visual Appearance of Foods to the Food Processor and the Consumer, *Journal of Food Quality* 1(3): 267–78.

Ihde, Don. 1990. *Technology and the Lifeworld: From Garden to Earth*. Bloomington, IN: Indiana University Press.

Ihde, Don. 2000. Epistemology Engines, *Nature* 406: 21.

Ihde, Don and Evan Selinger. 2004. Merleau-Ponty and Epistemology Engines, *Human Studies* 361–76.

Ilbery, Brian and Moya Kneafsey. 2000. Registering Regional Specialty Food and Drink Products in the United Kingdom: The Case of PDOs and PGIs, *Area* 32: 317–25.

Ingold, Tim. 1988. Introduction, pp. 1–16, in *What is an Animal?*, edited by Tim Ingold, London: Unwin Hyman.

Ingold, Tim. 1994. From Trust to Domination: An Alternative History of Human Animal Relations, in *Animals and Human Society: Changing Perspectives*, edited by A. Manning and J. Serpell, pp. 1–22, New York: Routledge.

Ingold, Tim. 1995. Building, Dwelling, Living: How People and Animals Make Themselves at Home in the World, in *Shifting Contexts Transformations in Anthropological Knowledge*, pp. 57–80, edited by M. Strathern, London: Routledge.

Ingold, Tim. 2000. *The Perception of the Environment: Essays on Livelihood, Dwelling, and Skill*. New York: Routledge.

Ingold, Tim. 2009. Point, Line and Counterpoint: From Environment to Fluid Space, in *Neurobiology of Umwelt: How Living Beings Perceive the World*, edited by A. Berthoz and Y. Christen, pp. 141–55, Berlin, Heidelberg: Springer-Verlag.

Ingram, Julie. 2008. Are Farmers in England Equipped to Meet the Knowledge Challenge of Sustainable Soil Management? An Analysis of Farmer and Advisor Views, *Journal of Environmental Management* 86(1): 214–28.

Irvine, Leslie. 2004. *If You Tame Me: Understanding Our Connection with Animals*. Philadelphia, PA: Temple University Press.

Jay, Martin. 1994. *Downcast Eyes: The Degeneration of Vision in Twentieth Century*. Berkeley, CA: University of California Press.

Jerolmack, Colin. 2007. Animal Practices, Ethnicity, and Community: The Turkish Pigeon Handlers of Berlin, *American Sociological Review* 72: 874–94.

Johnson, L.A., D.J. Myers, and D.J. Burden. 1992. Soy Protein's History, Prospects in Food, Feed, *International News on Fats, Oils, and Related Materials* 3(4): 429–44.

Johnston, Catherine. 2008. Beyond the Clearing: Towards a Dwelt Animal Geography, *Progress in Human Geography* 32(5): 633–49.

Johnston, Josee and Shyon Baumann 2010. *Foodies: Democracy and Distinction in the Gourmet Foodscape*. New York: Routledge.

Jonas, K., M. Diehl, and P. Bromer. 1997. Effects of Attitudinal Ambivalence on Information Processing and Attitude–Intention Consistency, *Journal of Experimental Social Psychology* 33: 190–210.

Jullien, Francois. 2002. Did Philosophers Have to Become Fixated on Truth?, *Critical Inquiry* Summer: 803–24.

Kains, M.G. 1910. *Chicken World: Profitable Poultry Production*. New York: Orange Judd Company.

Kauffman, Harold. 1999. *World Soybean Research Conference VI: Proceedings*. Minneapolis, MN: University of Minnesota Press.

Kellert, S. 1979. Zoological Parks in American Society, Address Given at the Meeting of American Association of Zoological Parks in Aquaria, St. Louis, MO.

Kellert, S. 1997. *Kinship to Mastery: Biophilia in Human Evolution and Development*. Washington, DC: Island Press.

King, Christine. 2008. Community Resilience and Contemporary Agri-Ecological Systems: Reconnecting People and Food, and People with People, *Systems Research and Behavioral Science* 25(1): 111–24.

Kirwan, James. 2006. The Interpersonal World of Direct Marketing: Examining Conventions of Quality at UK Farmers' Markets, *Journal of Rural Studies* 22: 301–12.

Kloppenburg, Jack. 1991. Social Theory and De/Reconstruction of Agricultural Science: For an Alternative Agriculture, *Rural Sociology* 56: 519–48.

Kloppenburg, Jack and Beth Burrows. 1996. Biotechnology to the Rescue? Twelve Reasons why Biotechnology is Incompatible with Sustainable Agriculture, *The Ecologist* 26(2): 61–7.

Kloppenburg, Jack and Charles Geisler. 1985. The Agricultural Ladder: Agrarian Ideology and the Changing Structure of U.S. Agriculture, *Journal of Rural Studies* 1: 59–72.

Kneafset, Moya, Rosie Cox, Lewis Holloway, Elizabeth Dowler, Laura Venn, and Helena Tuomainen. 2008. *Reconnecting Consumers, Producers and Food: Exploring Alternatives*. New York: Berg.

Knorr-Cetina, Karin. 1999. *Epistemic Cultures: How the Sciences Make Knowledge*. Cambridge, MA: Harvard University Press.

Kockelkoren, Petran. 2005. Art and Technology Playing Leapfrog: A History and Philosophy of Technoesis, in *Inside the Politics of Technology: Agency and Normativity in the Coproduction of Technology in Society*, edited by Hans Harbers, pp. 147–67, Amsterdam: Amsterdam University Press.

Korsmeyer, Carolyn. 1999. *Making Sense of Taste: Food and Philosophy*. Ithaca, NY: Cornell University Press.

Krasteva-Blagoeva, Evgenija. 2008. Tasting the Balkans: Food and Identity, *Ethnologia Balkanica* 12: 25–36.

Kress, Gunther and Theo Van Leeuwen. 1996. *Reading Images: The Grammar of Visual Design*. London: Routledge.

Kriflik, Lynda and Heather Yeatman. 2005. Food Scares and Sustainability: A Consumer Perspective, *Health, Risk and Society* 7(1): 11–24.

Kruse, Corwin. 1999. Gender, Views of Nature and Support for Animal Rights, *Society and Animals* 7(3): 179–98.

Kummer, Corby, Susie Cushner, Carlo Petrini, and Eric Schlosser. 2008. *The Pleasures of Slow Food: Celebrating Authentic Traditions, Flavors, and Recipes*. San Francisco, CA: Chronicle Books.

Laet, Marianne de. 2000. Patents, Travel, Space: Ethnographic Encounters with Objects in Transit, *Environment and Planning D: Society and Space* 18: 149–68.

Laet, Marianne de and Diederick Raven. 1989. Interview with Bruno Latour, *Focaal* 11/12: 193–208.

Lakoff, George and Mark Johnson. 1999. *Philosophy in the Flesh*. New York: Basic Books.

Lamb, Charles, Joseph Hair, Carl McDaniel. 2008. *Essentials of Marketing*. Cengage. Florence, KY.

Lang, Josephine. 2001. Managerial Concerns in Knowledge Management, *Journal of Knowledge Management* 5(1): 43–57.

Larmine, Claire. 2005. Settling Shared Uncertainties: Local Partnerships between Producers and Consumers, *Sociologia Ruralis* 45(4): 324–45.

Latham, Alan. 2003. Research, Performance, and Doing Human Geography: Some Reflections on the Diary-photograph, Diary-interview Method, *Environment and Planning A* 35: 1993–2017.

Latour, Bruno. 1987. *Science in Action: How to Follow Scientists and Engineers through Society*. Cambridge, MA: Harvard University Press.

Latour, Bruno. 1992. Where Are the Missing Masses? Sociology of a Few Mundane Artifacts, in, *Shaping Technology, Building Society: Studies in Sociotechnical Change*, edited by W. Bijker and J. Law, pp. 225–58, Cambridge, MA: MIT Press.

Latour, Bruno. 1993. *We Have Never Been Modern*. Cambridge, MA: Harvard University Press.

Latour, Bruno. 1999. *Pandora's Hope: Essays on the Reality of Science Studies.* Cambridge, MA: Harvard University Press.

Latour, Bruno. 2004a. How to Talk About the Body: The Normative Dimension of Science Studies, *Body and Society* 10 (2–3): 2–5–229.

Latour, Bruno. 2004b. Why has Critique Run Out of Steam? From Matters of Fact to Matters of Concern, *Critical Inquiry* 30(winter): 225–48.

Lave, Jean and Etienne Wenger. 1991. *Situated Learning: Legitimate Peripheral Participation*. New York: Cambridge University Press.

Law, John. 2002. *Aircraft Stories: Decentering the Object in Technoscience*. Durham, NC: Duke University Press.

Law, John and Annemarie Mol. 2008. Globalisation in Practice: On the Politics of Boiling Pigswill, *Geoforum* 39: 133–43.

Lee, Roger. 2006. The Ordinary Economy: Tangled Up in Values and Geography, *Transitions of the Institute of British Geographers* 31(4): 413–32.

Lee, Sandra S.-J. 2000. Dys-appearing Tongues and Bodily Memories: The Aging of First-Generation Resident Koreans in Japan, *Ethos* 28: 198–223.

Levi-Strauss, Claude. 1963. *Totemism*. London: Beacon Press.

Levine, Susan. 2008. *School Lunch Politics: The Surprising History of America's Favorite Welfare Program*. Princeton, NJ: Princeton University Press.

Lieberman, Leslie. 2006. Evolutionary and Anthropological Perspectives on Optimal Foraging in Obesogenic Environments, *Appetite* 47(1): 3–9.

Lockie, Stewart. 2002. The Invisible Mouth: Mobilizing the Consumer in Food Production-Consumption Networks, *Sociologia Ruralis* 42(2): 278–94.

Lockie, Stewart and Simon Kitto. 2000. Beyond the Farm Gate: Production-Consumption Networks and Agri-Food Research, *Sociologia Ruralis* 40(1): 3–19.

Longhurst, Robyn, Lynda Johnston, and Elsie Ho. 2009. A Visceral Approach: Cooking "at Home" with Migrant Women in Hamilton, New Zealand, *Transactions Institute of British Geography* 34: 333–45.

Lorimer, Hayden. 2005. Cultural Geography: The Busyness of Being 'More-Than-Representational', *Progress in Human Geography* 29(1): 83–94.

Lupton, Deborah. 1996. *Food, the Body, and the Self*. Thousand Oaks, CA: Sage.

Lyson, Thomas. 1979. Going to College: An Emerging Rung on the Agricultural Ladder, *Rural Sociology* 44: 773–90.

Lyson, Thomas. 2004. *Civic Agriculture: Reconnecting Farm, Food, and Community*. Medford, MA: Tufts University Press.

MacKenzie, Adrian. 2002. *Transductions*. London: Continuum.

MacKenzie, Donald and Graham Spinardi. 1995. Tacit Knowledge, Weapons Design, and the Uninvention of Nuclear Weapons, *American Journal of Sociology* 101(1): 44–99.

Maddox. J. 1991. Sad Tale of an Endangered Zoo, *Nature* 350: 457.

Magdoff, Fred, John Foster, and Frederick Buttel. 2000. *Hungry for Profit: The Agribusiness Threat to Farmers, Food, and the Environment*. New York: Monthly Review Press.

Mann, Susan and James Dickenson. 1978. Obstacles to the Development of a Capitalist Agriculture, *Journal of Peasant Studies* 5(4): 466–81.

Mansfield, Becky. 2003. Fish, Factory Trawlers, and Imitation Crab: The Nature of Quality in the Seafood Industry, *Journal of Rural Studies* 19(1): 9–21.

Mansfield, Becky and Johanna Haas. 2006. Scale Framing of Scientific Uncertainty in Controversy Over the Endangered Steller Sea Lion, *Environmental Politics* 15(1): 78–94.

Mansvelt, Juliana. 2009. Geographies of Consumption: Engaging with Absent Presences, *Progress in Human Geography* 1: 1–10.

Marston, Sallie. 2000. The Social Construction of Scale, *Progress in Human Geography* 24(2): 219–42.

Martinez, D. 1996. First People, Firsthand Knowledge, *Sierra* Nov/Dec: 50–51, 70–71.

McAfee, Kathleen. 2003. Neoliberalism on the Molecular Scale, *Geoforum* 34(2): 203–21.

McCloud, J.T. 1974. Soy Protein in School Feeding Programs, *Journal of the American Chemists' Society* 51(1): 141A–142A.

McCormack, Derek. 2002. A Paper with an Interest in Rhythm, *Geoforum* 33: 469–85.

McCullum, C., E. Desjardins, V. Kraak, P. Ladipo, and H. Costello. 2005. Evidence-Based Strategies to Build Community Food Security, *Journal of the American Dietetic Association* 105(2): 278–83.

McEwan, Chris and Michael Goodman. 2010. Place, Geography and the Ethics of Care: Introductory Remarks on the Geographies of Ethics, Responsibility, and Care, *Ethics, Place and Environment* 13(2): 103–12.

McNeely, Jeffery and Sara Siebert. 2002. *Ecoagriculture: Strategies to Feed the World and Save Wild Biodiversity*. Washington, DC: Island Press.

McWilliams, James. 2005. *A Revolution in Eating: How the Quest for Food Shaped America*. New York: Columbia University Press.

Mead, George Herbert. 1934. *Mind, Self and Society.* Chicago, IL: University of Chicago Press.

Medical and Surgical Reporter. 1879. Notes and Comments: Siderodromophobia, *The Medical and Surgical Reporter: A Weekly Journal* 40(1164): 529–50.

Merleau-Ponty, Maurice. 1992 (1945). *The Phenomenology of Perception*. New York: Routledge.

Mentreddy, S.R., A.I. Mohamed, N. Joshee, and A.K. Yadav. 2002. Edamame: A Nutritious Vegetable Crop, in *Trends in New Crops and New Uses*, edited by J. Janick and A. Whipkey, pp. 432–8, Alexandria, VA: ASHS Press.

Michaels, David and Celeste Monforton. 2005. Manufacturing Uncertainty: Contested Science and the Protection of the Public's Health and Environment, *American Journal of Public Health* 95: S39–48.

Miele, Mara and Jonathan Murdoch. 2002. The Practical Aesthetics of Traditional Cuisines: Slow Food in Tuscany, *Sociologia Ruralis* 42(4): 312–28.

Milestad, Rebecka, Ruth Bartel-Kratochvil, Heidrun Leitner, and Paul Axmann. 2010. Being Close: The Quality of Social Relationships in a Local Organic Cereal and Bread Network in Lower Austria, *Journal of Rural Studies* 26(3): 228–40.

Miller, Steve. 2001. Public Understanding of Science at the Crossroads, *Public Understanding of Science* 10: 115–20.

Mills, C. Wright. 2000. *The Sociological Imagination*. Oxford: Oxford University Press.

Mintrom, Michael. 1997. Policy Entrepreneurs and the Diffusion of Innovation, *American Journal of Political Science* 41(3): 738–70.

Mintz, Sidney. 1996. *Tasting Food, Tasting Freedom: Excursions into Eating, Culture, and the Past*. Boston, MA: Beacon Press.

Mintz, Sidney. 2002. Food and Eating: Some Persisting Questions, in *Food Nations: Selling Taste in Consumer Societies*, edited by Warren Belasco and Philip Scranton, pp. 24–33, New York: Routledge.

Mintz, Sidney, Chee-Beng Tan, and Christine DuBois. 2008. Introduction: The Significance of Soy, in *The World of Soy*, edited by Christine DuBois, Chee-Beng Tan, and Sidney Mintz, pp. 1–26, Urbana, IL: University of Illinois Press.

Morgan, K. 2004. The Exaggerated Death of Geography: Learning, Proximity and Territorial Innovation Systems, *Journal of Economic Geography* 4: 3–21.

Morrison, Toni. 2008. *What Moves at the Margin: Selected Nonfiction*. Jackson, MS: University Press of Mississippi.

Mort, Maggie. 2002. *Building the Trident Network: A Study of the Enrollment of People, Knowledge, and Machines*. Cambridge, MA: MIT Press.

Murphy, Sophia. 2008. Globalization and Corporate Concentration in the Food and Agricultural Sector, *Development* 51(4): 527–33.

Myin, Erik. 2003. An Account of Color Without a Subject? *Behavioral and Brain Sciences* 26: 42–3.

Nabhan, Gary. 1997. *Cultures of Habitat*. Washington, DC: Counterpoint.

Nagel, Thomas. 1970. *The Possibility of Altruism*. Princeton, NJ: Princeton University Press.

Nazarea, Virginia. 1998. *Cultural Memory and Biodiversity*. Tucson, AZ: The University of Arizona Press.

Nazarea, Virginia. 2005. *Heirloom Seeds and Their Keepers: Marginality and Memory in the Conservation of Biological Diversity*. Tucson, AZ: The University of Arizona Press.

Nestle, Marion. 2003. *Safe Food: Bacteria, Biotechnology, and Bioterrorism*. Berkeley, CA: University of California Press.

Nestle, Marion. 2010. Food Politics Blog, April 16, http://www.foodpolitics.com/2010/04/can-pepsico-help-alleviate-world-hunger, last accessed August 15, 2010.

Niebuhr, Reinhold. 1934. *Moral Man and Immoral Society: A Study in Ethics and Politics*. New York: Charles Scribner's Sons.

Nietzsche, Fredrick. 1969. *Twilight of the Idols*. London: Henry, and New York; Macmillan.

Norretranders, Tor. 1998. *The User Illusion*. New York: Viking.

Ogilvie, Brian. 2006. *The Science of Describing: Natural History in Renaissance Europe*. Chicago, IL: University of Chicago Press.

Okin, Susan. 1989. Reason and Feeling in Thinking About Justice, *Ethics* 99: 229–49.

Ong, Walter. 2004. *Orality and Literacy: The Technologizing of the Word*. New York: Routledge.

Oosterveer, Peter. 2007. *Global Governance of Food Production and Consumption: Issues and Challenges*. Northampton, MA: Edward Elgar Publishing.

Orr, Julian. 1996. *Talking About Machines: An Ethnography of a Modern Job*. Ithaca, NY: Cornell University Press.

Orwell, George. 1992. *1984*. New York: Random House.

Owen, James. 2005. Medieval Lion Skulls Reveal Secrets of Tower of London "Zoo", *National Geographic*, November 3 http://news.nationalgeographic.com/news/2005/11/1103_051103_tower_lions.html, last accessed August 1, 2010.

Owen-Smith, Jason and Walter Powell. 2004. Knowledge Networks as Channels and Conduits: The Effects of Spillovers in the Boston Biotechnology Community, *Organization Science* 15(1): 5–21.

Pearce, Susan. 1995. *On Collecting: An Investigation into Collecting in the European Tradition*. London: Routledge.

Perrow, Charles. 1999. *Normal Accidents: Living with High Risk Technologies*. Princeton, NJ: Princeton University Press.

Peterson, Anna. 2009. *Everyday Ethics and Social Change: The Education of Desire*. New York: Columbia University Press.

Peterson, Franklynn. 1974. The Bean that's Making Meat Obsolete, *Popular Mechanics* 142(4): 84–7, 188.

Peterson, Wesley. 2009. *A Billion Dollars a Day: The Economics and Politics of Agricultural Subsides*. Malden, MA: Wiley-Blackwell.

Petrini, Carol and Benjamin Watson. 2001. *Slow Food: Collected Thoughts on Taste, Tradition, and the Honest Pleasures of Food*. White River, VT: Chelsea Green.

Petrini, Carlo and William McCuaig. 2004. *Slow Food: The Case for Taste*. New York: Columbia University Press.

Philo, Chris. 1995. Animals, Geography, and the City: Notes on Inclusions and Exclusions, *Environment and Planning D: Society and Space* 13, pp. 655–81.

Philpott, Tom. 2007. The Short Term Solution that Stuck: Where Farm Subsidies Came From and Why They're Still Here, *Grist* January 30, http://www.grist.org/comments/food/2007/01/30/farm_bill2/, last accessed December 13, 2009.

Pifer, Linda, Shimizu Kinya, and Ralph Pifer. 1994. Public Attitudes Toward Animal Research: Some International Comparisons, *Society and Animals* 2(2): 95–113.

Poerksen, Bernhard. 2006. Truth is What Works: Francisco J. Varela on Cognitive Science, Buddhism, the Inseparability of Subject and Object, and the Exaggerations of Constructivism—A Conversation, *Journal of Aesthetic Education* 40(1) 35–53.

Polakowski, J. 1987. *Zoo Design: The Reality of Wild Illusions*. Ann Arbor, MI: University of Michigan Press.

Polanyi, Karl. 2001 (1944). *The Great Transformation: The Political and Economic Origins of Our Time*. New York: Beacon Press.

Polanyi, Michael. 1966. *The Tacit Dimension*. Garden City, NY: Doubleday.

Pollan, Michael. 2001. *Botany of Desire: A Plant's Eye View of the World*. New York: Random House.

Pollan, Michael. 2008. *In Defense of Food: An Eater's Manifesto*. New York: Penguin Group.

Ponds, Roderik, Frank van Oort, and Koen Frenken. 2010. Innovation, Spillovers and University-Industry Collaboration: An Extended Knowledge Production Function Approach, *Journal of Economic Geography* 10(2): 231–55.

Povey, R., B. Wellens, and M. Conner. 2001. Attitudes Towards Following Meat, Vegetarian and Vegan Diets: An Examination of the Role of Ambivalence, *Appetite* 37: 15–26.

Powell, Horace. 1956. *The Original Has this Signature: WK Kellogg*. Englewood Cliffs, NJ: Prentice Hall.

Powell, Walter, Kenneth Koput, and Laurel Smith-Doerr. 1996. Interorganizational Collaboration and the Locus of Innovation: Networks of Learning in Biotechnology, *Administrative Science Quarterly* 41(1): 116–45.

Princen, Thomas. 1997. The Shading and Distancing of Commerce: When Internalization is Not Enough, *Ecological Economics* 20: 235–53.

Pringle, Peter. 2005. *Food Inc.: Mendel to Monsanto—The Promises and Perils of the Biotech Harvest*. New York: Simon and Schuster.

Probyn, Elspeth. 1999. Beyond Food/Sex: Eating and the Ethics of Resistance, *Theory, Culture and Society* 16(2): 215–28.

Probyn, Elspeth. 2001. *Carnal Appetites: Food, Sex, Identities*. New York: Taylor and Francis.

Proctor, Robert. 1996. *Cancer Wars: How Politics Shapes What We Know and Don't Know*. New York: Basic Books.

Proctor, Robert. 2006. Everyone Knew But No One Had "Proof": Tobacco Industry Use of Medical History Expertise in US Courts, 1990–2002, *Tobacco Control* 15: iv117–25.

Raynolds, Laura. 2002. Consumer/Producer Links in Fair Trade Coffee Networks, *Sociologia Ruralis* 42(4): 404–24.

Reeds, Karen. 1991. *Botany in Medieval and Renaissance Universities*. New York: Garland.

Renard, Marie-Christine. 2003. Fair Trade: Quality, Market and Conventions, *Journal of Rural Studies* 19: 87–96.

Renfrew, Jane, Maggie Black, Pete Bears, and Jennifer Stead. 1985. *Food and Cooking in 19th Century Britain, History and Recipes*. London: English Heritage.

Revill, George. 2004. Cultural Geographies in Practice: Performing French Folk Music: Dance, Authenticity and Nonrepresentational Theory, *Cultural Geographies* 11(2): 199–209.

Rigney, Ann. 2005. Plentitude, Scarcity, and the Circulation of Cultural Memory, *Journal of European Studies* 35(1): 11–28.

Ritzer, George. 2003. Islands of the Living Dead: The Social Geography of McDonaldization, *American Behavioral Scientist* 47(2): 119–36.

Ritzer, George. 2008. *McDonaldization of Society, 5th edition*. Thousand Oaks, CA: Sage.

Robinson, T., M. Saphir, H. Kraemer, A. Varady, and K. Haydel. 2001. Effects of Reducing Television Viewing on Children's Requests for Toys, *Developmental and Behavioral Pediatrics* 229(3): 179–84.

Robinson, T., D. Brozekowski, D. Matheson, and H. Kraemer. 2007. Effects of Fast Food Branding on Young Children's Taste Preferences, *Archive of Pediatrics and Adolescent Medicine* 161(8): 792–7.

Rodaway, Paul. 1994. *Sensuous Geographies*. New York: Routledge.

Roe, Emma. 2006. Things Becoming Food and the Embodied, Material Practices of an Organic Food Consumer, *Sociologia Ruralis* 46(2): 104–21.

Rogers, Everett. 1988. *Social Change in Rural Societies: An Introduction to Rural Sociology*. Upper Saddle River, NJ: Prentice Hall.

Rogers, Everett. 1995. *Diffusion of Innovations*. New York: Simon and Schuster.

Röling, Niels and Elske van de Fliert. 1994. Transforming Extension for Sustainable Agriculture: The Case for Integrated Pest Management in Rice in Indonesia, *Agriculture and Human Values* 11(2–3): 96–108.

Rosa, Deborah. 1996. *Nourishing Terrains: Australian Aboriginal Views of Landscape and Wilderness*. Australia: Canberra.

Rose, Gillian. 2001. *Visual Methodologies: An Introduction to the Interpretation of Visual Materials*. London: Sage.

Rosenthal, Sandra. 1986. *Speculative Pragmatism*. Amherst, MA: University of Massachusetts Press.

Ross, Nancy. 2006. How Civic is It? Success Stories in Locally Focused Agriculture in Maine, *Renewable Agriculture and Food Systems* 21(2): 114–23.

Ryan, Bryce and Neal Gross. 1943. The Diffusion of Hybrid Seed Corn in Two Iowa Communities, *Rural Sociology* 8: 15–24.

Schoonover, Heather and Mark Muller. 2006. Food Without Thought: How US Farm Policy Contributes to Obesity, Institute for Agriculture and Trade Policy, Environment and Agriculture Program, March, http://www.iatp.org/iatp/factsheets.cfm?accountID=258&refID=89968, last accessed August 15, 2010.

Scott, James. 1999. *Seeing Like a State: How Certain Schemes to Improve the Human Condition Have Failed.* New Haven, CT: Yale University Press.

Selfa, Theresa and Joan Qazi. 2005. Place, Taste, or Face-to-Face? Understanding Producer-Consumer Networks in "Local" Food Systems in Washington State, *Agriculture and Human Values* 22: 451–64.

Seremetakis, C. Nadia. 1993. The Memory of the Senses: Historical Perception, Communal Exchange and Modernity, *Visual Anthropological Review* 9(2): 2–18.

Shackley, Simon, Peter Young, Stuart Parkinson, and Brian Wynne. 1998. Uncertainty, Complexity and Concepts of Good Science in Climate Change Modelling: Are GCMs the Best Tools?, *Climate Change* 38(2): 159–205.

Shapiro, Kenneth. 1997. A Phenomenological Approach to the Study of Nonhuman Animals, in *Anthropomorphism, Anecdotes and Animals*, edited by R. Mitchell, N. Thompson, and H. Miles, Albany, pp. 277–95, New York: State University of New York Press.

Sheasley, Bob. 2008. *Home to Roost: A Backyard Farmer Chases Chickens Through the Ages.* New York: Macmillan.

Shiva, Vandana. 1993. *Monocultures of the Mind: Perspectives on Biodiversity and Biotechnology.* New York: Palgrave Macmillan.

Shteyngart, Gary. 2006. *Absurdistan.* New York: Random House.

Shurtleff, William and Akiko Aoyagi. 2001. *The Book of Tempeh.* Lafayette, CA: Soyinfo Center.

Sibley, David. 1995. *Geographies of Exclusion: Society and Difference in the West.* London: Routledge.

Singer, Peter. 1995. *How Are We to Live? Ethics in an Age of Self-Interest.* Amherst, NY: Prometheus Books.

Slocum, Rachel. 2008. Thinking Race Through Corporeal Feminist Theory: Divisions and Intimacies at the Minneapolis Farmers' Market, *Social and Cultural Geography* 9: 849–69.

Smith, Adam. 1976. *The Theory of Moral Sentiments.* Oxford: Clarendon Press.

Smith, Adrian. 2006. Green Niches in Sustainable Development: The Case of Organic Food in the United Kingdom, *Environment and Planning C* 24(3): 439–58.

Smith, Mick. 2002. The "Ethical" Space of the Abattoir: On the (In)human(e) Slaughter of Other Animals, *Human Ecology Review* 9(2): 49–58.

Smith, Neil. 1992. Contours of a Spatialized Politics: Homeless Vehicles and the Production of Geographic Scale, *Social Text* 33: 55–81.

Soo-jin Lee, Sandra. 2000. Dys-Appearaing Tongues and Body Memories: The Aging of First-Generation Resident Koreans in Japan, *Ethos* 28(2): 198–223.

Sousa de, Ivan, Sergio Freire and Lawrence Busch. 1998. Networks and Agricultural Development: The Case of Soybean Production and Consumption in Brazil, *Rural Sociology* 63(3): 349–71.

Sparks, P., M. Conner, R. James, R. Sheperd, and R. Povey. 2001. Ambivalence About Health-Related Behaviors: An Exploration in the Domain of Food Choice, *British Journal of Health Psychology* 6: 53–68.

Spillman, Lyn and Brian Conway. 2007. Texts, Bodies, and the Memory of Bloody Sunday, *Symbolic Interaction* 30(1): 79–103.

Spillman, W.J. 1930. The Agricultural Ladder, in *A Systematic Source Book in Rural Sociology*, edited by P.A. Sorokin, C.C. Zimmerman, and C. Galpin, pp. 523–28, New York: Russell and Russell.

Spitz, Peter. 1988. *Petrochemicals: The Rise of an Industry*. New York: John Wiley.

Squier, Susan. 2006. Chicken Auguries, *Configurations* 14: 69–86.

Stanley, Herbert. 1920. *Bygone Beliefs: Being a Series of Excursions in the Bylaws of Thought*. London: William Rider and Son.

Stengers, Isabelle. 2002. A Cosmo-Politics: Risk, Hope, and Change, in *Hope: New Philosophies for Change*, edited by Mary Zournazi, pp. 244–72, New York: Routledge.

Stoller, Paul. 1989. *The Taste of Ethnographic Things: The Senses in Anthropology*. Philadelphia, PA: University of Pennsylvania Press.

Strange, Marty. 1988. *Family Farming: A New Economic Vision*. Lincoln, NE: University of Nebraska Press.

Strauss, Anselm. 1959. *Mirrors and Masks: The Search for Identity*. Glenco, IL: Free Press.

Stuart, Tristram. 2009. *Waste: Uncovering the Global Food Scandal*. New York: Norton.

Sutter, Cindy. 2009. Soy Crazy: The US Consumes Tons of Soy—and It's Not All Good, *Daily Camera* November 24, http://www.dailycamera.com/food/ci_13835378#axzz0gkeZsjg5, last accessed February 27, 2010.

Sutton, David. 2001. *Remembrance of Repasts: An Anthropology of Food and Memory*. London: Berg.

Swabe, Joanna, Bart Rutgers, and Elsbeth Noordhuizen-Stassen. 2005. Cultural Attitudes Towards Killing Animals, in *The Human-Animal Relationship: Forever and a Day*, pp. 123–39, edited by Francien Henriëtte de Jonge and Ruud van den Bos, Assen, The Netherlands: Uitgeverij Van Gorcum.

Taras, H., J. Sallis, T. Patterson, P. Nader, and J. Nelson. 1989. Television's Influence on Children's Diet and Physical Activity, *Journal of Developmental and Behavioral Pediatrics* 10: 176–80.

Tegtmeier, Erin and Michael Duffy. 2004. External Costs of Agricultural Productivity in the United States, *International Journal of Agricultural Sustainability* 2: 1–20.

Terdiman, Richard. 2003. Given Memory: On Mnemonic Coercion, Reproduction and Invention, in *Regimes of Memory*, edited by S. Radstone and K. Hodgkin, pp. 186–201, London: Routledge.

Thien, Deborah. 2005. After or Beyond Feeling? A Consideration of Affect and Emotion in Geography, *Area* 37 (4): 450–56.

Thrift, Nigel. 1998. Steps to an Ecology of Place, in *Human Geography Today*, edited by D. Massey, J. Allen, A. Sarre, pp. 295–323, Cambridge: Polity.

Thrift, Nigel. 2000. Still Life in Nearly Present Time, *Body and Society* 6: 34–57.

Thrift, Nigel. 2004. Summoning Life, in *Envisioning Human Geographies*, edited by P. Cloke, P. Crang, and M. Goodwin, pp. 81–103, New York: Edward Arnold.

Thrift, Nigel. 2005. From Born to Made: Technology, Biology and Space, *Transactions of the Institute of British Geographers* 30(4): 463–76.

Thrupp, Lori. 2000. Linking Agricultural Biodiversity and Food Security: The Valuable Role of Sustainable Agriculture, *International Affairs* 76(2): 265–81.

Toland, Dan. 2009. Eda-What-e? Farmer Works to Bring Ohio-grown Edamame Mainstream, *Our Ohio* November/December, http://ourohio.org/magazine/past-issues-2009/november-december-2009/edamame/, last accessed March 1, 2010.

Trubek, Amy. 2008. *The Taste of Place: A Cultural Journey into the Terroir*. Los Angeles, CA: University of California Press.

Tuan, Y.-F. 1984. *Dominance and Affection: The Making of Pets*. London: University Press.

Tucker, M., S. Whaley, and J. Sharp. 2006. Consumer Perceptions of Food-Related Risks, *International Journal of Food Science and Technology* 41(2): 135–46.

Tulloch, John and Deborah Lupton. 2002. Consuming Risk, Consuming Science, *Journal of Consumer Culture* 2(3): 363–83.

Turkley, Sherry. 2007. *Evocative Objects: Things We Think With*. Cambridge, MA: MIT Press.

Turkley, Sherry. 2008a. *Falling for Science: Objects in Mind*. Cambridge, MA: MIT Press.

Turkley, Sherry. 2008b. *The Inner History of Devices*. Cambridge, MA: MIT Press.

Turkley, Sherry. 2009. *Simulation and Its Discontents*. Cambridge, MA: MIT Press.

Turner, Simon. 2010. Networks of Learning within the English Wine Industry, *Journal of Economic Geography* 10(5): 685–715.

Unruh, Gregory. 2000. Understanding Carbon Lock In, *Energy Policy* 28: 817–30.

USDA (United States Department of Agriculture). 2004. *2002 Census of Agriculture* (National Agricultural Statistics Service, United States Department of Agriculture, Washington, DC).

USDA (United States Department of Agriculture). 2007. *The Census of Agriculture, 2007*. National Agricultural Statistics Service, Washington, DC, http://www. agcensus.usda.gov/Publications/2007/index.asp, last accessed September 1, 2010.

Vannini, Philip, Guppy Ahluwalia-Lopez, Dennis Waskul, and Simon Gottschalk. 2010. Performing Taste at Wine Festivals: A Somatic Layered Account of Material Culture, *Qualitative Inquiry* 16(5): 378–96.

Varela, Francisco. 1999. *Ethical Know-How*. Stanford, CA: Stanford University Press.

Varela, Francisco, Evan Thompson, and Eleanor Rosch. 1992. *The Embodied Mind: Cognitive Science and Human Experience*. Boston, MA: MIT Press.

Vileisis, Ann. 2007. *Kitchen Literacy: How We Lost Knowledge of Where Food Comes From and Why We Need to Get it Back*. Washington, DC: Island Press.

Wainwright, Steven. 1995. The Transformational Experience of Liver Transplantation, *Journal of Advanced Nursing* 22: 1068–76.

Wainwright, Steven and Bryan Turner. 2004. Epiphanies of Embodiment: Injury, Identity and the Balletic Body, *Qualitative Research* 4(3): 311–37.

Wansink, Brian. 2007. *Mindless Eating: Why We Eat More Than We Think*. New York: Random House.

Warde, Alan and Lydia Martens. 1999. *Eating Out, Social Differentiation, Consumption and Pleasure*. Cambridge: Cambridge University Press.

Waters, Alice. 2008. *Art of Simple Food*. London: Penguin UK.

Watts, D., B. Ilbery, and D. Maye. 2005. Making Reconnections in Agro-Food Geography: Alternative Systems of Food Provision, *Progress in Human Geography* 29(1): 22–40.

Wecox, A.G. 1885. *The Dixie Cookbook*. Atlanta, GA: L.A. Clarkson and Company.

Weis, Tony. 2007. *The Global Food Economy: The Battle for the Future of Farming*. New York: Zed Books.

Weiss, Allen. 1998. *Unnatural Horizons: Paradox and Contradiction in Landscape Architecture*. Princeton, NJ: Princeton Architectural Press.

Wenger, Etienne. 1998. *Communities of Practice: Learning, Meaning, and Identity*. New York: Cambridge University Press.

Wenger, Etienne, Richard McDermott, and William Snyder. 2002. *Cultivating Communities of Practice*. Cambridge, MA: Harvard University Press.

Whatmore, Sarah. 2002. *Hybrid Geographies*. London: Sage.

Whitehead, A.N. 1967 (1933). *Adventures of Ideas*. New York: Free Press.

Whitehead, A.N. 1938. *Modes of Thought.* Cambridge: Cambridge University Press.

Wiecha J., K. Peterson, D. Ludwig, J. Kim, A. Sobol, and S. Gortmaker. 2006. When Children Eat What They Watch: Impact of Television Viewing on Dietary Intake in Youth, *Archives of Pediatric and Adolescent Medicine* 160(4): 436–42.

Wik, Reynold. 1962. Henry Ford's Science and Technology for Rural America, *Technology and Culture* 3(3): 247–58.

Williamson, Judith. 1983. *Decoding Advertisements: Ideology and Meaning in Advertising*. London: Marion Boyars.

Wilson, Geoff. 2001. From Productivism to Post-Productivism ... and Back Again? Exploring the (Un)changed Natural and Mental Landscapes of European Agriculture, *Transactions of the Institute of British Geographers* 26(1): 77–102.

Winders, Bill. 2009. *The Politics of Food Supply: US Agricultural Policy in the World Economy*. New Haven, CT: Yale University Press.

Winter, Michael. 1997. New Policies and New Skills: Agricultural Change and Technology Transfer, *Sociologia Ruralis* 37(3): 363–81.

Winter, Michael. 2003. Embeddedness, the New Food Economy and Defensive Localism, *Journal of Rural Studies* 19(1): 23–32.

Wittgenstein, Ludwig. 1969. *On Certainty*. Oxford: Basil Blackwell.

Wolin, Sheldon. 1997. What Time is It? *Theory and Event* 1:1.

Wood, David. 2005. *The Step Back*. New York: State University of New York Press.

Worley, Stephen. 2009. Where the Buffalo Roamed, *WeatherSealed.com*, http://www.weathersealed.com/2009/09/22/where-the-buffalo-roamed, last accessed September 30, 2009.

Wright, Christopher and John Lund. 2003. Supply Chain Rationalization: Retailer Dominance and Labour Flexibility in the Australian Food and Grocery Industry, *Work, Employment and Society* 17(1): 137–57.

Wright, Wynne and Gerad Middendorf. 2008. *The Fight Over Food: Producers, Consumers, and Activists Challenge the Global Food System*. College Station, PA: Pennsylvania State Press.

Wynne, Brian. 1995. Public Understanding of Science, in *Handbook of Science and Technology*, edited by S. Jasanoff, G. Markle, J. Petersen, and T. Pinch, pp. 361–88, Thousand Oaks, CA: Sage.

Yarwood, Richard and Nick Evans. 2000. Taking Stock of Farm Animals and Rurality, in *Animal Spaces, Beastly Places*, edited by C. Philo and C. Wilbert, pp. 98–114, New York: Routledge.

Yates, Frances. 1966. *The Art of Memory*. Chicago, IL: University of Chicago Press.

Yeong, Choy. 2009. Dean says US Soy Milk Sales may Reach $1 Billion, *Bloomberg News* June 15, http://www.bloomberg.com/apps/news?pid=20601081&sid=aX1Je8Kr7S6U, last accessed February 27, 2010.

Zwart, Hub. 1997. What is an Animal? A Philosophical Reflection on the Possibility of a Moral Relationship with Animals, *Environmental Values* 6: 377–92.

Index